PUBLICATIONS OF
INTERNATIONAL S[...]
(INSTITUT UNIVER[...] DE HAUTES ETUDES
INTERNATIONALES, GENÈVE, SUISSE)

No. 18

MONETARY NATIONALISM

AND

INTERNATIONAL STABILITY

BY

F. A. von HAYEK

Tooke Professor of Economic
Science and Statistics in the University
of London

Martino Publishing
Mansfield Centre, CT
2012

Martino Publishing
P.O. Box 373,
Mansfield Centre, CT 06250 USA

www.martinopublishing.com

ISBN 978 1 61427 341-7

© 2012 Martino Publishing

All rights reserved. No new contribution to this publication may
be reproduced, stored in a retrieval system, or transmitted, in any form or
by any means, electronic, mechanical, photocopying, recording, or otherwise,
without the prior permission of the Publisher.

Cover design by T. Matarazzo

Printed in the United States of America On 100% Acid-Free Paper

PUBLICATIONS OF THE GRADUATE INSTITUTE OF
INTERNATIONAL STUDIES, GENEVA, SWITZERLAND
(INSTITUT UNIVERSITAIRE DE HAUTES ETUDES
INTERNATIONALES, GENÈVE, SUISSE)

No. 18

MONETARY NATIONALISM
AND
INTERNATIONAL STABILITY

BY

F. A. von HAYEK

Tooke Professor of Economic
Science and Statistics in the University
of London

LONGMANS, GREEN AND CO
LONDON - NEW YORK - TORONTO
1937

LONGMANS, GREEN AND CO. LTD.
39 PATERNOSTER ROW LONDON, E.C.4
6 OLD COURT HOUSE STREET, CALCUTTA
53 NICOL ROAD, BOMBAY
36A MOUNT ROAD, MADRAS

LONGMANS, GREEN AND CO.
114 FIFTH AVENUE, NEW YORK
221 EAST 20TH STREET, CHICAGO
88 TREMONT STREET, BOSTON

LONGMANS, GREEN AND CO.
480 UNIVERSITY AVENUE, TORONTO

Printed in Belgium. All rights reserved

NOTE

SINCE its establishment in 1927, the Graduate Institute of International Studies at Geneva has organised, in addition to its regular instruction, certain series of lectures by experts from outside its own staff. In spite of the original character of most of these lectures, the Institute cannot publish all of them.

The Directors believe, however, that they ought to make certain of them available to a larger public.

The task of the Directors of the Institute is limited to the selection of their authors. The authors have been left complete freedom of thought, and naturally, therefore, they bear the sole responsibility not only for their opinions, but also for the form of their expression.

Paul MANTOUX.
W. E. RAPPARD.

PREFACE

THE five lectures which are here reproduced are necessarily confined to certain aspects of the wide subject indicated by the title. They are printed essentially as they were delivered and, as is explained in the first lecture, limitations of time made it necessary to chose between discussing the concrete problems of the present policy of Monetary Nationalism and concentrating on the broader theoretical issues on which the decision between an international standard and independent national currencies must ultimately be based. The first course would have involved a discussion of such technical questions as the operations of Exchange Equalization Accounts, Forward Exchanges, the choice and adjustment of parities, cooperation between central banks, etc., etc. The reader will find little on these subjects in the following pages. It appeared to me more important to use the time available to discuss the general ideas which are mainly responsible for the rise of Monetary Nationalism and to which it is mainly due that policies and pratices which not long ago would have been frowned upon by all responsible financial experts, are now generally employed throughout the world. The immediate influence of the theoretical speculation is probably weak, but that it has had a profound influence in shaping those views which to-day dominate monetary policy is not open to serious question. It seemed to me better therefore to concentrate on these wider issues.

This decision has permitted me a certain freedom in the discussion of alternative policies. In discussing the merits of various systems I have not felt bound to confine myself to those which may to-day be considered practical politics. I have no doubt that to those who take the present trend of intellectual development for granted much of the discussion in the following pages will appear highly academic. Yet fundamentally the alternative policies here considered are no more revolutionary or impracticable than the deviations from traditional practice which have been widely discussed and which have even been attempted in recent years—except that at the moment not so many people believe in them. But while the politician—and the economist when he is advising on concrete measures—must take the state of opinion for granted in deciding what changes can be contemplated here and now, these limitations are not necessary when we are asking what is best for the human race in general. I am profoundly convinced that it is academic discussion of this sort which in the long run forms public opinion and which in consequence decides what will be practical politics some time hence. I regard it therefore not only as the privilege but as the duty of the academic economist to take all alternatives into consideration, however remote their realisation may appear at the moment.

And indeed I must confess that it seems to me in many respects the future development of professional and public opinion on these matters is much more important than any concrete measure which may be taken in the near future. Whatever the permanent arrangements in monetary policy, the spirit in which the existing institutions are administered is at least as important as these institutions themselves. And just as, long before the breakdown of the international gold standard in 1931, monetary policy all over the world was guided by the ideas of Monetary Nationalism which eventually brought its breakdown, so at the present time there is grave danger that a restoration of the external apparatus of the gold standard may not mean a return to a really international currency. Indeed I must admit that—although I am a convinced believer in the international gold standard—I regard the prospects of its restoration in the near future not without some concern. Nothing would be more fatal from a long run point of view than if the world attempted a formal return to the gold standard before people had become willing to work it, and if, as would be quite probable under these circumstances, this were soon followed by a renewed collapse. And although this would probably be denied by the advocates of Monetary Nationalism, it seems to me as if we had reached a stage where their views have got such a hold on those in responsible positions, where so much of the traditional rules of policy have either been forgotten or been displaced by others which are, unconsciously perhaps, part of the new philosophy, that much must be done in the realm of ideas before we can hope to achieve the basis of a stable international system. These lectures were intended as a small contribution to this preparatory work which must precede a successful reconstruction of such a system.

It was my good fortune to be asked to deliver these lectures at the *Institut Universitaire de Hautes Etudes Internationales* at Geneva. I wish here to express my profound gratitude for the opportunity thus afforded and for the sympathetic and stimulating discussion which followed the lectures. My thanks are particularly due to the directors of the Institute, Professors Rappard and Mantoux, not only for arranging the lectures but also for undertaking their publication in the present series.

I am also indebted to a number of my friends and colleagues at the London School of Economics, particularly to Dr. F. Benham, Mr. F. Paish, Professor Robbins, and Mr. C. H. Secord, who have read the manuscript and offered much valuable advice as regards the subject matter and the form of exposition of these lectures. This would certainly have been a much bigger and much better book if I had seen my way to adopt and incorporate all their suggestions. But at the moment I do not feel prepared to undertake the larger investigation which my friends rightly think the subject deserves. I alone must therefore bear the blame for the sketchy treatment of some important points and for any shortcomings which offend the reader.

I hope however it will be born in mind that these lectures were written to be read aloud and that this forbade any too extensive discussion of the more intricate theoretical points involved. Only at a few points, I have added a further explanatory paragraph or restored sections which would not fit into the time available for the lecture. That this will not suffice to provide satisfactory answers to the many questions I have raised I have no doubt.

F.A. VON HAYEK.

*London School of Economics
and Political Science.*

May 1937.

CONTENTS

LECTURE I. *National Monetary Systems*   1
   1. Theoretical character of these lectures, p. 1;
   2. Monetary Nationalism and national monetary systems, p. 4; 3. A homogeneous international currency, p. 5; 4. The mixed or national reserve system, p. 8; 5. Independent national currencies, p. 14.

LECTURE II. *The Function and Mechanism of International Flows of Money* 17
   1. The functions of redistributions of the world's stock of money, p. 17; 2. The mechanism under a system of purely metallic currencies, p. 19; 3. The mechanism under a regime of mixed currencies, p. 25; 4. The rôle of central bank policy, p. 32.

LECTURE III. *Independent Currencies* 35
   1. National stabilisation and international shifts of demand, p. 35; 2. The position in the country adversely affected, p. 38; 3. The position in the country favourably affected, p. 41; 4. Causes of the recent growth of Monetary Nationalism, p. 43; 5. The significance of the concepts of inflation and deflation applied to a national area, p. 47.

LECTURE IV. *International Capital Movements* 54
   1. Definition and classification of capital movements, p. 54; 2. Their mechanism under a homogeneous international currency, p. 57; 3. Under national reserve systems, p. 60; 4. Under independent national currencies, p. 63; 5. The control of international capital movements, p. 68.

LECTURE V. *The Problems of a Really International Standard* 73
   1. Gold as the international standard, p. 73; 2. The " perverse elasticity " of our credit money, p. 76; 3. The Chicago Plan of banking reform, p. 81; 4. Exchange rates and gold reserves, p. 84; 5. Credit policy of the central banks, p. 88; 6. Conclusions, p. 92.

Lecture I

NATIONAL MONETARY SYSTEMS

1

When I was honoured with the invitation to deliver at the *Institut* five lectures " on some subject of distinctly international interest ", I could have little doubt what that subject should be. In a field in which I am particularly interested I had been watching for years with increasing apprehension the steady growth of a doctrine which, if it becomes dominant, is likely to deal a fatal blow to the hopes of a revival of international economic relations. This doctrine, which in the title of these lectures I have described as Monetary Nationalism, is held by some of the most brilliant and influential economists of our time. It has been practised in recent years to an ever increasing extent, and in my opinion it is largely responsible for the particular intensification of the last depression which was brought about by the successive breakdown of the different currency systems. It will almost certainly continue to gain influence for some time to come, and it will probably indefinitely postpone the restoration of a truly international currency system. Even if it does not prevent the restoration of an international gold standard, it will almost inevitably bring about its renewed breakdown soon after it has been re-established.

When I say this I do not mean to suggest that a restoration of the gold standard of the type we have known is necessarily desirable, nor that much of the criticism directed against it may not be justified. My complaint is rather that most of this criticism is not concerned with the true reasons why the gold standard, in the form in which we knew it, did not fulfill the functions for which it was designed; and further that the only alternatives which are seriously considered and discussed, completely abandon what seems to me the essentially sound principle—that of an international currency system—which that standard is supposed to embody.

But let me say at once that when I describe the doctrines I am going to criticize as Monetary Nationalism I do not mean to suggest that those who hold them are actuated by any sort of narrow nationalism. The very name of their leading exponent, Mr. J. M. Keynes, testifies that this is not the case. It is not the motives which inspire those who advocate such plans, but the consequences which I believe would follow from their realization, which I have in mind when I use this term. I have no doubt that the advocates of these doctrines sincerely believe that the system of independent national currencies will reduce rather than increase the causes of international economic friction; and that not merely one country but all will in the long run be better off if there is established that freedom in national monetary policies which is incompatible with a single international monetary system.

The difference then is not one about the ultimate ends to be achieved. Indeed, if it were, it would be useless to try to solve it by rational discussion. The fact is rather that there are genuine differences of opinion among economists about the consequences of the different types of monetary arrangements we shall have to consider, differences which prove that there must be inherent in the problem serious intellectual difficulties which have not yet been fully overcome. This means that any discussion of the issues involved will have to grapple with considerable technical difficulties, and that it will have to grapple with wide problems of general theory if it is to contribute anything to their solution. My aim throughout will be to throw some light on a very practical and topical problem. But I am afraid my way will have to lead for a considerable distance through the arid regions of abstract theory.

There is indeed another way in which I might have dealt with my subject. And when I realized how much purely theoretical argument the other involved I was strongly tempted to take it. It would have been to avoid any discussion of the underlying ideas and simply to take one of the many concrete proposals for independent national currency systems now prevalent and to consider its various probable effects. I have no doubt that in this form I could give my lectures a much more realistic appearance and could prove to the satisfaction of all who have already an unfavourable opinion of Monetary Nationalism that its effects are pernicious. But I am afraid I would have had little chance of convincing anyone who has already been attracted by the other side of the case. He might even admit all the disadvantages of the proposal which I could enumerate, and yet believe that its advantages outweigh the defects. Unless I can show that these supposed advantages are largely illusory, I shall not have got very far. But this involves an examination of the argument of the other side. So I have come rather reluctantly to the conclusion that I cannot shirk the much more laborious task of trying to go to the root of the theoretical differences.

2

But it is time for me to define more exactly what I mean by Monetary Nationalism and its opposite, an International Monetary System. By Monetary Nationalism I mean the doctrine that a country's share in the world's supply of money should *not* be left to be determined by the same principles and the same mechanism as those which determine the relative amounts of money in its different regions or localities. A truly International Monetary System would be one where the whole world possessed a homogeneous currency such as obtains within separate countries and where its flow between regions was left to be determined by the results of the action of all individuals. I shall have to define later what exactly I mean by a homogeneous currency. But I should like to make it clear at the outset that I do not believe that the gold standard as we knew it conformed to that ideal and that I regard this as its main defect.

Now from this conception of Monetary Nationalism there at once arises a question. The monetary relations between small adjoining areas are alleged to differ from those between larger regions or countries; and this difference is supposed to justify or demand different monetary arrangements. We are at once led to ask what is the nature of this alleged difference? This question is somewhat connected but not identical with the question what constitutes a national monetary system, in what sense we can speak of different monetary systems. But, as we shall see, it is very necessary to keep these questions apart. For if we do not we shall be confused between differences which are inherent in the underlying situation and which may make different monetary arrangements desirable, and differences which are the consequence of the particular monetary arrangements which are actually in existence.

For reasons which I shall presently explain this distinction has not always been observed. This has led to much argument at cross purposes, and it is therefore necessary to be rather pedantic about it.

3

I shall begin by considering a situation where there is as little difference as is conceivable between the money of different countries, a case indeed where there is so little difference that it becomes doubtful whether we can speak of different " systems ". I shall assume two countries only and I shall assume that in each of the two countries of which our world is assumed to consist, there is only one sort of widely used medium of exchange, namely coins consisting of the same metal. It is irrelevant for our purpose whether the denomination of these coins in the two countries is the same, so long as we assume, as we shall, that the two sorts of coins are freely and without cost interchangeable at the mints. It is clear that the mere difference in denomination, although it may mean an inconvenience, does not constitute a relevant difference in the currency systems of the two countries.[1]

[1] Since these lines were written a newly published book has come to my hand in which almost the whole argument in favour of Monetary Nationalism is based on the assumption that different national currencies are different commodities and that consequently there ought to be variable prices of them in terms of each other. (C. R. WITTHLESEY, *International Monetary Issues*, New York, 1937.) No attempt is made to explain why or under what conditions and in what sense the different national moneys ought to be regarded as different commodities, and one can hardly avoid the impression that the author has uncritically accepted the difference of denomination as proof of the existence of a difference in kind. The case illustrates beautifully the prevalent confusion between differences between the currency systems which can be made an argument for national differentiations and those which are a

In starting from this case we follow a long established precedent. A great part of the argument of the classical writers on money proceeded on this assumption of a " purely metallic currency ". I wholly agree with these writers that for certain purposes it is a very useful assumption to make. I shall however not follow them in their practice of assuming that the conclusions arrived from these assumptions can be applied immediately to the monetary systems actually in existence. This belief was due to their conviction that the existing mixed currency systems not only could and should be made to behave in every respect in the same way as a purely metallic currency, but that—at any rate in England since the Bank Act of 1844—the total quantity of money was *actually* made to behave in this way. I shall argue later that this erroneous belief is responsible for much confusion about the mechanism of the gold standard as it existed; that it has prevented us from achieving a satisfactory theory of the working of the modern mixed system, since the explanation of the rôle of the banking system was only imperfectly grafted upon, and never really integrated with, the theory of the purely metallic currency; and that in consequence the gold standard or the existence of an international system was blamed for much which in fact

---

consequence of such differentiations. That is " only a difference in nomenclature " (as Professor Gregory has well put it) whether we express a given quantity of gold as Pounds, Dollars or Marks, and that this no more constitutes different commodities than the same quantity of cloth becomes a different commodity when it is expressed in meters instead of in yards, ought to be obvious. Whether different national currencies are in any sense different commodities depends on what we make them, and the real problem is whether we should create differentiations between the national currencies by using in each national territory a kind of money which will be generally acceptable only within that territory, or whether the same money should be used in the different national territories.

NATIONAL MONETARY SYSTEMS

was really due to the mixed character of the system and not to its " internationalism " at all.

For my present purpose, however, namely to find whether and in what sense the monetary mechanism of one country can or must be regarded as a unit or a separate system, even when there is a minimum of difference between the kind of money used there and elsewhere, the case of the " purely metallic currency " serves extraordinarily well. If there are differences in the working of the national monetary systems which are not merely an effect of the differences in the monetary arrangements of different countries, but which make it desirable that there *should* be separate arrangements for different regions, they must manifest themselves even in this simplest case.

It is clear that in this case the argument for a national monetary system cannot rest on any peculiarities of the national money. It must rest, and indeed it does rest, on the assumption that there is a particularly close connection between the prices—and particularly the wages—within the country which causes them to move to a considerable degree up and down together compared with prices outside the country. This is frequently regarded as sufficient reason why, in order to avoid the necessity that the " country as a whole " should have to raise or lower its prices, the quantity of money in the country should be so adjusted as to keep the " general price level " within the country stable. I do not want to consider this argument yet. I shall later argue that it rests largely on an illusion, based on the accident that the statistical measures of prices movements are usually constructed for countries as such; and that in so far as there are genuine difficulties connected with general downward adjustments of many prices, and particularly wages, the proposed remedy would be worse than the disease. But I think I ought to say here and now that I regard it as the only argument on which the case for monetary nationalism can be rationally based. All the other arguments have really nothing to do with the existence of an international monetary system as such, but apply only to the particular sorts of international systems with which we are familiar. But since these arguments are so inextricably mixed up in current discussion with those of a more fundamental character it becomes necessary, before we can consider the main arguments on its merits, to consider them first.

The homogeneous international monetary system which we have just considered was characterised by the fact that each unit of the circulating medium of each country could equally be used for payments in the other country and for this purpose could be bodily transferred into that other country and be bodily transformed int the currency of that country. Among the systems which need to be considered only an international gold standard with exclusive gold circulation in all countries would conform to this picture. This has never existed in its pure form and the type of gold standard which existed until fairly recently was even further removed from this picture than was generally realized. It was never fully appreciated how much the operation of the system which actually existed diverged from the ideal pure gold standard. For the points of divergence were so familiar that they were usually taken for granted. It was the design of the Bank Act of 1844 to make the mixed system of gold and other money behave in such a way that the quantity of money would change exactly as if only gold were in circulation; and for a long time argument proceeded as if this intention had actually been realized. And even when it was

NATIONAL MONETARY SYSTEMS

gradually realized that deposits subject to cheque were no less money than bank notes, and that since they were left out of the regulation, the purpose of the Act had really been defeated, only a few modifications of the argument were thought necessary. Indeed in general this argument is still presented as it was originally constructed, on the assumption of a purely metallic currency.

In fact however with the coming of modern banks a complete change had occurred. There was no longer one homogeneous sort of money in each country, the different units of which could be regarded as equivalent for all relevant purposes. There had arisen a hierarchy of different kinds of money within each country, a complex organisation which possessed a definite structure, and which is what we really mean when we speak of the circulating medium of a country as a " system ". It is probably much truer to say that it is the difference between the different kinds of money which are used in any one country, rather than the differences between the moneys used in different countries, which constitutes the real difference between different monetary systems.

We can see this if we examine matters a little more closely. The gradual growth of banking habits, that is the practice of keeping liquid assets in the form of bank balances subject to cheque, meant that increasing numbers of people were satisfied to hold a form of the circulating medium which could be used directly only for payments to people who banked with the same institution. For all payments beyond this circle they relied on the ability of the bank to convert the deposits on demand into another sort of money which was acceptable in wider circles; and for this purpose the banks had to keep a " reserve " of this more widely acceptable or more liquid medium.

But this distinction between bank deposits and " cash "

in the narrower sense of the term does not yet exhaust the classification of different sorts of money, possessing different degrees of liquidity, which are actually used in a modern community. Indeed, this development would have made little difference if the banks themselves had not developed in a way which led to their organisation into banking " systems " on national lines. Whether there existed only a system of comparatively small local unit banks, or whether there were numerous systems of branch banks which covered different areas freely overlapping and without respect to national boundaries, there would be no reason why all the monetary transactions within a country should be more closely knit together than those in different countries. For any excess payments outside their circle the customers of any single bank, it is true, would be dependent on the reserve kept for this purpose for them by their bank, and might therefore find that their individual position might be affected by what other members of this circle did. But at most the inhabitants of some small town would in this way become dependent on the same reserves and thereby on one another's action,[1] never all the inhabitants of a big area or a country.

It was only with the growth of centralized national banking systems that all the inhabitants of a country came in this sense to be dependent on the same amount of more liquid assets held for them collectively as a national reserve. But the concept of centralisation in this connection must not be interpreted too narrowly as referring only to systems crowned by a central bank of the familiar type, nor even as confined to branch banking systems where each district of a country is served by the branches of the same few banks. The forms in which centralisa-

[1] Compare on this and the following L. ROBBINS, *Economic Planning and International Order*, 1937, pp. 274 *et seq.*

NATIONAL MONETARY SYSTEMS 11

tion, in the sense of a system of national reserves which is significant here, may develop, are more varied than this and they are only partly due to deliberate legislative interference. They are partly due to less obvious institutonal factors.

For even in the absence of a central bank and of branch banking the fact that a country usually has one financial centre where the stock exchange is located and through which a great proportion of its foreign trade passes or is financed tends to have the effect that the banks in that centre become the holders of a large part of the reserve of all the other banks in the country. The proximity of the stock exchange puts them in a position to invest such reserves profitably in what, at any rate for any single bank, appears to be a highly liquid form. And the greater volume of transactions in foreign exchange in such a centre makes it natural that the banks outside will rely on their town correspondents to provide them with whatever foreign money they may need in the course of their business. It was in this way that long before the creation of the Federal Reserve System in 1913 and in spite of the absence of branch banking there developed in the United States a system of national reserves under which in effect all the banks throughout their territory relied largely on the same ultimate reserves. And a somewhat similar situation existed in Great Britain before the growth of joint stock banking.

But this tendency is considerably strengthened if instead of a system of small unit banks there are a few large joint stock banks with many branches; still more if the whole system is crowned by a single central bank, the holder of the ultimate cash reserve. This system, which to-day is universal, means in effect that additional distinctions of acceptability or liquidity have been artificially created between three main types of money, and that the task of keeping a sufficient part of the total assets in liquid form for different purposes has been divided between different subjects. The ordinary individual will hold only a sort of money which can be used directly only for payments to clients of the same bank; he relies upon the assumption that his bank will hold for all its clients a reserve which can be used for other payments. The commercial banks in turn will only hold reserves of such more liquid or more widely acceptable sort of money as can be used for inter-bank payments within the country. But for the holding of reserves of the kind which can be used for payments abroad, or even those which are required if the public should want to convert a considerable part of its deposits into cash, the banks rely largely on the central bank.

This complex structure, which is often described as the one-reserve system, but which I should prefer to call the system of national reserves, is now taken so much for granted that we have almost forgotten to think about its consequences. Its effects on the mechanism of international flows of money will be one of the main subjects of my next lecture. To-day I only want to stress two aspects which are often overlooked. In the first place I would emphasize that bank deposits could never have assumed their present predominant rôle among the different media of circulation, that the balances held on current account by banks could never have grown to ten times and more of their cash reserves, unless some organ, be it a privileged central bank or be it a number of or all the banks, had been put in a position, to create in case of need a sufficient number of additional bank notes to satisfy any desire on the part of the public to convert a considerable part of their balances into hand-to-hand money. It is in this sense and in this sense only that the existence of a national reserve system involves the question of the regulation of the note issue alone.

The second point is that nearly all the practical problems of banking policy, nearly all the questions with which a central banker is daily concerned, arise out of the co-existence of these different sorts of money within the national monetary system. Theoretical economists frequently argue as if the quantity of money in the country were a perfectly homogeneous magnitude and entirely subject to deliberate control by the central monetary authority. This assumption has been the source of much mutual misunderstanding on both sides. And it has had the effect that the fundamental dilemma of all central banking policy has hardly ever been really faced : the only effective means by which a central bank can control an expansion of the generally used media of circulation is by making it clear in advance that it will not provide the cash (in the narrower sense) which will be required in consequence of such expansion, but at the same time it is recognised as the paramount duty of a central bank to provide that cash once the expansion of bank deposits has actually occurred and the public begins to demand that they should be converted into notes or gold.

I shall be returning to this problem later. But in the next two lectures my main concern will be another set of problems. I shall argue that the existence of national reserve systems is the real source of most of the difficulties which are usually attributed to the existence of an international standard. I shall argue that these difficulties are really due to the fact that the mixed national currencies are not sufficiently international, and that most of the criticism directed against the gold standard *qua* international standard is misdirected. I shall try to show that the existence of national reserve systems alters the mechanism of the international money flows from what it would be with a homogeneous international currency to a much greater degree than is commonly realized.

5

But before I can proceed to this major task I must shortly consider the third and most efficient cause which may differentiate the circulating media of different countries and constitute separate monetary systems. Up to this point I have only mentioned cases where the ratio between the monetary units used in the different countries was given and constant. In the first case this was secured by the fact that the money circulating in the different countries was assumed to be homogeneous in all essential respects, while in the second and more realistic case it was assumed that, although different kinds of money were used in the different countries, there was yet in operation an effective if somewhat complicated mechanism which made it always possible to convert at a constant rate money of the one country into money of the other. To complete the list there must be added the case where these ratios are variable : that is, where the rate of exchange between the two currencies is subject to fluctuations.

With monetary systems of this kind we have of course to deal with differences between the various sorts of money which are much bigger than any we have yet encountered. The possession of a quantity of money current in one country no longer gives command over a definite quantity of money which can be used in another country. There is no longer a mechanism which secures that an attempt to transfer money from country to country will lead to a decrease in the quantity of money in one country and a corresponding increase in the other. In

NATIONAL MONETARY SYSTEMS 15

fact an actual transfer of money from country to country becomes useless because what is money in the one country is not money in the other. We have here to deal with things which possess different degrees of usefulness for different purposes and the quantities of which are fixed independently.

Now I think it should be sufficiently clear that any differences between merely interlocal and international movements of money which only arise as a consequence of the variability of exchange rates cannot themselves be regarded as a justification for the existence of separate monetary systems. That would be to confuse effect and cause—to make the occasion of difference the justification of its perpetuation. But since the adoption of such a system of " flexible parities " is strongly advocated as a remedy for the difficulties which arise out of other differences which we have already considered, it will be expedient if in the following lectures I consider side by side all three types of conditions under which differences between the national monetary systems may arise. We shall be concerned with the way in which in each case redistributions of the relative amounts of money in the different countries are effected. I shall begin with the only case which can truly be described as an international monetary standard, that of a homogeneous international currency. Consideration of this case will help me to show what functions changes in the relative quantities of money in different regions and countries may be conceived to serve; and how such changes are spontaneously brought about. I shall then proceed to the hybrid " mixed " system which until recently was the system generally in vogue and which is meant when, in current discussion, the traditional gold standard is referred to. As I said at the beginning, I shall not deny that this system has serious defects. But while the Monetary Nationalists believe that these defects are due to the fact that it is still an international system and propose to remove them by substituting the third or purely national type of monetary system for it, I shall on the contrary attempt to show that its defects lie in the impediments which it presents to the free international flow of funds. This will then lead me first to an examination of the peculiar theory of inflation and deflation on which Monetary Nationalism is based : then to an investigation of the consequences which we should have to expect if its proposals were acted upon; and finally to a consideration of the methods by which a more truly international system could be achieved.

Lecture II

THE FUNCTION AND MECHANISM OF INTERNATIONAL FLOWS OF MONEY

1

At the end of my first lecture I pointed out that the three different types of national monetary systems which we have been considering differed mainly in the method by which they effected international redistributions of money. In the case of a homogeneous international currency such a redistribution is effected by actual transfers of the corresponding amounts of money from country to country. Under the " mixed " system represented by the traditional gold standard—better called " gold nucleus standard "—it is brought about partly by an actual transfer of money from country to country; but largely by a contraction of the credit super-structure in the one country and a corresponding expansion in the other. But although the mechanism and, as we shall see, some of the effects, of these two methods are different, the final result, the change in the relative value of the total quantities of money in the different countries, is brought about by a corresponding change in the quantity of money, the number of money units, in each country. Under the third system, however, the system of independent currencies, things are different. Here the adjustment is brought about, not by a change in the number of money units in each country, but by changes in their relative value. No money actually passes from country to country, and whatever redistribution of money between persons may be involved by the redistribution between countries has to be brought about by corresponding changes inside each country.

Before, however, we can assess the merits of the different systems it is necessary to consider generally the different reasons why it may become necessary that the relative values of the total quantities of money in different countries should alter. It is clear that changes in the demand for or supply of the goods and services produced in an area may change the value of the share of the world's income which the inhabitants of that area may claim. But changes in the relative stock of money, although of course closely connected with these changes of the shares in the world's income which different countries can claim, are not identical with them. It is only because people whose money receipts fall will in general tend to reduce their money holdings also and *vice-versa*, that changes in the size of the money stream in the different countries will as a rule be accompagnied by changes in the same direction in the size of the money holdings. People who find their income increasing will generally at first take out part of the increased money income in the form of a permanent increase in their cash balances, while people whose incomes decrease will tend to postpone for a while a reduction of their expenditure to the full extent, preferring to reduce their cash balances.[1] To this extent changes in the cash balances serve, as it were, as cushions which soften the impact and delay the adaptation of the real incomes to the changed money incomes, so that in the interval money is actually taken as a substitute for goods.

But, given existing habits, it is clear that changes in the relative size of money incomes—and the same applies

---

[1] For a full description of this mechanism cf. R. G. HAWTREY, *Currency and Credit*, chapter IV, 3rd ed., 1928, pp. 41-63.

to the total volume of money transactions—of different countries make corresponding changes in the money stocks of these countries inevitable; changes which, although they need not be in the same proportion, must at any rate be in the same direction as the changes in incomes. If the share in the world's production which the output of a country represents, rises or falls, the share of the total which the inhabitants of the country can claim will fully adapt itself to the new situation only after money balances have been adjusted.[1]

Changes in the demand for money on the part of a particular country may of course also occur independently of any change in the value of the resources its inhabitants can command. They may be due to the fact that some circumstances may have made its people want to hold a larger or smaller proportion of their resources in the most liquid form, *i.e.* in money. If so, then for a time they will offer to the rest of the world more commodities, receiving money in exchange. This enables them, at any later date, to buy more commodities than they can currently sell. In effect they decide to lend to the rest of the world that amount of money's worth of commodities in order to be able to call it back whenever they want it.

2

The function which is performed by international movements of money will be seen more clearly if we proceed to consider such movement in the simplest case

---

[1] Perhaps, intead of speaking of the world's output, I should have spoken about the share in the command over the world's resources, since of course it is not only the current consumable product but equally the command over resources which will yield a product only in the future which is distributed by this monetary mechanism.

imaginable—a homogeneous international or "purely metallic" currency. Let us suppose that somebody who used to spend certain sums on products of country A now spends them on products of country B. The immediate effect of this is the same whether this person himself is domiciled in A or in B. In either case there will arise an excess of payments from A to B—an adverse balance of trade for A—, either because the total of such payments has risen or because the amount of payments in the opposite direction has fallen off. And if the initiator of this change persists in his new spending habits, this flow of money will continue for some time.

But now we must notice that because of this in A somebody's money receipts have decreased and in B somebody's money receipts have increased. We have long been familiar with the proposition that counteracting forces will in time bring the flow of money between the countries to a stop. But it is only quite recently that the exact circumstances determining the route by which this comes about have been satisfactorily established[1]. In both countries the change in the money receipts of the people first affected will be passed on and disseminated. But how long the outflow of money from A to B will continue depends on how long it takes before the successive changes in money incomes set up in each country will bring about new and opposite changes in the balance of payments.

This result can be brought about in two ways in each of the two countries. The reduction of money incomes in country A may lead to a decrease of purchases from B,

---

[1] Cf. particularly F. W. PAISH, *Banking Policy and the Balance of International Payments* (*Economica*, N. S., vol. III, no. 12, Nov. 1936); K. F. MAIER, *Goldwanderungen*, Jena 1935, and P. B. WHALE, *The Working of the Pre-War Gold Standard* (*Economica*, vol. IV, no. 13, February 1937).

INTERNATIONAL FLOWS OF MONEY 21

or the consequent fall of the prices of some goods in A may lead to an increase of exports to B. And the increase of money incomes in country B may lead to an increase of purchases from A or to a rise in the prices of some commodities in B and a consequent decrease of exports to A. But how long it will take before in this way the flow of money from A to B will be offset will depend on the number of links in the chains which ultimately lead back to the other country, and on the extent to which at each of these points the change of incomes leads first to a change in the cash balances held, before it is passed on in full strength. In the interval money will continue to flow from A to B; and the total which so moves will correspond exactly to the amounts by which, in the course of the process just described, cash balances have been depleted in the one country and increased in the other.

This part of the description is completely general. But we cannot say how many incomes will have to be changed, how many individual prices will have to be altered upwards or downwards in each of the two countries, in consequence of the initial changes. For this depends entirely on the concrete circumstances of each particular case. In some countries and under some conditions the route will be short because some of the first people whose incomes decrease cut down their expenditure on imported goods, or because the increase of incomes is soon spent on imported goods.[1] In other cases the route may be long and external payments will be made to balance only after extensive price changes have occured, which induce further people to change the direction of their expenditure.

The important point in all this is that what incomes

---

[1] Cf. on this particularly the article by F. W. Paish just quoted.

and what prices will have to be altered in consequence of the initial change will depend on whether and to what extent the value of a particular factor or service, directly or indirectly, depends on the particular change in demand which has occurred, and not on whether it is inside or outside the same " currency area ". We can see this more clearly if we picture the series of successive changes of money incomes, which will follow on the initial shift of demand, as single chains, neglecting for the moment the successive ramifications which will occur at every link. Such a chain may either very soon lead to the other country or first run through a great many links at home. But whether any particular individual in the country will be affected will depend whether he is a link in that particular chain, that is whether he has more or less immediately been serving the individuals whose income has first been affected, and not simply on whether he is in the same country or not. In fact this picture of the chain makes it clear that is is not impossible that most of the people who ultimately suffer a decrease of income in consequence of the initial transfer of demand from A to B may be in B and not in A. This is often overlooked because the whole process is presented as if the chain of effects came to an end as soon as payments between the two countries balance. In fact however each of the two chains—that started by the decrease of somebody's income in A, and that started by the increase of another persons income in B—may continue to run on for a long time after they have passed into the other country, and may have even a greater number of links in that country than in the one where they started. They will come to an end only when they meet, not only in the same country but in the same individual, so finally offsetting each other. This means that the number of reductions of individual incomes and prices (not their aggregate amount) which becomes necessary in consequence of a transfer of money from A to B may actually be greater in B than in A.

This picture is of course highly unrealistic because it leaves out of account the infinite ramifications to which each of these chains of effects will develop. But even so it should, I think, make it clear how superficial and misleading the kind of argument is which runs in terms of *the* prices and *the* incomes of the country, as if they would necessarily move in unison or even in the same direction. It will be prices and incomes of particular individuals and particular industries which will be affected and the effects will not be essentially different from those which will follow any shifts of demand between different industries or localities.

This whole question is of course the same as that which I discussed in my first lecture in connection with the problem of what constitutes one monetary system, namely the question of whether there exists a particularly close coherence between prices and incomes, and particularly wages, in any one country which tends to make them move as a whole relatively to the price structure outside. As I indicated then, I shall not be able to deal with it more completely until later on. But there are two points which, I think, will have become clear now and which are important for the understanding of the contrast between the working of the homogeneous international currency we are considering, and the mixed system to which I shall presently proceed.

In the first place it already appears very doubtful whether there is any sense in which the terms inflation and deflation can be appropriately applied to these interregional or international transfers of money. If, of course, we *define* inflation and deflation as changes in the quantity of money, or the price level, *within a parti-*

cular territory, then the term naturally applies. But it is by no means clear that the consequences which we can show will follow if the quantity of money in a closed system changes will also apply to such redistributions of money between areas. In particular there is no reason why the changes in the quantity of money within an area should bring about those merely temporary changes in relative prices which, in the case of a real inflation, lead to misdirections of production—misdirections because eventually the inherent mechanism of these inflations tends to reverse these changes in relative prices. Nor does there seem any reason why, to use a more modern yet already obsolete terminology, saving and investment should be made to be equal within any particular area which is part of a larger economic system.[1] But all these questions can be really answered only when I come to discuss the two conflicting views about the main significance of inflation and deflation which underlie most of the current disputes about monetary policy.

The second point which I want particularly to stress here is that with a homogeneous international currency there is apparently no reason why an outflow of money from one area and an inflow into another should necessarily cause a rise in the rate of interest in the first area and a fall in the second. So far I have not mentioned the rate of interest, because there seems to be no general ground why we should expect that the causes which lead to the money flows between two countries should affect the rate of interest one way or the other. Whether they will have such an effect and in what direction will depend entirely on the concrete circumstances. If the initial change which reduces the money income of some people in one country leads to an immediate reduction of their expenditure on consumers' goods, and if in addition they

---

[1] Cf. J. M. KEYNES, *A Treatise on Money*, 1930, vol. I, chapter 4.

INTERNATIONAL FLOWS OF MONEY 25

use for additional investments the surplus of their cash balances which they no longer regard worth keeping, it is not impossible that the effect may actually be a fall in the rate of interest.[1] And, conversely, in the country towards whose product an additional money stream is directed, this might very well lead to a rise in the rate of interest. It seems that we have been led to regard what happens to be the rule under the existing mixed systems as due to causes much more fundamental than those which actually operate. But this leads me to the most important difference between the cases of a " purely metallic " and that of a " mixed " currency. To the latter case, therefore, I now turn.

3

If in the two countries concerned there are two separate banking systems, whether these banking systems are complete with a central bank or not, considerable transfers of money from the one country to the other will be effected by the actual transmission of only a part of the total, the further adjustment being brought about by an

---

[1] Although it is even conceivable that a fall in incomes might bring about a temporary rise in investments, because the people who are now poorer feel that they can no longer afford the luxury of the larger cash balances they used to keep before, and proceed to invest part of them, this is neither a very probable effect nor likely to be quantitatively significant. Much more important, however, may be the effect of the fall of incomes on the demand for investment. Particularly if the greater part of the existing capital equipment is of a very durable character a fall in incomes may for some time almost completely suspend the need for investment and in this way reduce the rate of interest in the country quite considerably. Another case where the same cause which would lead to a flow of money from one country to another would at the same time cause a fall in the rate of interest in the first would be if in one of several countries where population used to increase at the same rate, this rate were considerably decreased.

expansion or contraction of the credit structure according as circumstances demand. It is commonly believed that nothing fundamentally is changed but something is saved by substituting the extinction of money in one region and the creation of new money in the other for the actual transfer of money from individual to individual. This is however a view which can be held only on the most mechanistic form of the quantity theory and which completely disregards the fact that the incidence of the change will be very different in the two cases. Considering the methods available to the banking system to bring about an expansion or contraction, there is no reason to assume that they can take the money to be extinguished exactly from those persons where it would in the course of time be released if there were no banking system, or that they will place the additional money in the hands of those who would absorb the money if it came to the country by direct transfer from abroad. There are on the contrary strong grounds for believing that the burden of the change will fall entirely, and to an extent which is in no way justified by the underlying change in the real situation, on investment activity in both countries.

To see why and how this will happen it is necessary to consider in some detail the actual organisation of the banking systems and the nature of their traditional policies. We have seen that where bank deposits are used extensively this means that all those who hold their most liquid assets in this form, rely on their banks to provide them whenever needed with the kind of money which is acceptable outside the circle of the clients of the bank. The banks in turn, and largely because they have learnt to rely on the assistance of other (note issuing) banks, particularly the central bank, have come themselves to keep only very slender cash reserves, that is, reserves

INTERNATIONAL FLOWS OF MONEY 27

which they can use to meet any adverse clearing balance to other banks or to make payments abroad. These are indeed not meant to do more than to tide over any temporary and relatively small difference between payments and receipts. They are altogether insufficient to allow the banks ever to reduce these reserves by the full amount of any considerable reduction of their deposits. The very system of proportional reserves, which so far as deposits are concerned is to-day universally adopted and even in the case of bank notes applies practically everywhere outside Great Britain, means that the cash required for the conversion of an appreciable part of the deposits has to be raised by compelling people to repay loans.

We shall best see the significance of such a banking structure with respect to international money flows if we consider again the effects which are caused by an initial transfer of demand from country A to country B. The main point here is that, with a national banking system working on the proportional reserve principle, unless the adverse balance of payments corrects itself very rapidly, the central bank will not be in a position to let the outflow of money go on until it comes to its natural end. It cannot, without endangering its reserve position, freely convert all the bank deposits or banknotes which will be released by individuals into money which can be transferred into the other countries. If it wants to prevent an exhaustion or dangerous depletion of its reserves it has to speed up the process by which payments from A to B will be decreased or payments from B to A will be increased. And the only way in which it can do this quickly and effectively is generally and indiscriminately, to bring pressure on those who have borrowed from it to repay their loans. In this way it will set up additional chains of successive reductions of outlay, first on the part of those to whom it would have lent and then on the part of all others to whom this money would gradually have passed. So that leaving aside for the moment the effects which a rise in interest rates will have on international movements of short term capital we can see that the forces which earlier or later will reduce payments abroad and, by reducing prices of home products, stimulate purchases from abroad will be intensified. And if sufficient pressure is exercised in this way, the period during which the outflow of money continues, and thereby the total amount of money that will actually leave the country before payments in and out will balance again, may be reduced to almost any extent.

The important point, however, is that in this case the people who will have to reduce their expenditure in order to produce that result, will not necessarily be the same people who would ultimately have to do so under a homogeneous international currency system, and that the equilibrium so reached will of its nature be only temporary. In particular, since bank loans, to any significant extent, are only made for investment purposes, it will mean that the full force of the reduction of the money stream will have to fall on investment activity. This is shown clearly by the method by which this restriction is brought about. We have seen before that under a purely metallic currency an outflow of money need not actually bring about a rise in interest rates. It may, but it is not necessary and it is even conceivable that the opposite will happen. But with a banking structure organised on national lines, that is, under a national reserve system, it is inevitable that it will bring a rise in interest rates, irrespective of whether the underlying real change has affected either the profitability of investment or the rate of savings in such a way as to justify such a change. In other words, to use an expression which has given rise to much dispute in the recent past but which should be readily understood in this connection, the rise of the bank rate under such circumstances means that it has to be deliberately raised above the equilibrium or " natural " rate of interest.[1] The reason for this is not, or need not be, that the initiating change has affected the relation between the supply of investible funds and the demand for them, but that it tends to disturb the customary proportion between the different parts of the credit structure and that the only way to restore these proportions is to cancel loans made for investment purposes.

To some extent, but only to some extent, the credit contraction will, as I have just said, by lowering prices induce additional payments from abroad and in this form offset the outflow of money. But to a considerable extent its effect will be that certain international transfers of money which would have taken the place of a transfer of goods and would in this sense have been a final payment for a temporary excess of imports will be intercepted, so that consequently actual transfers of goods will have to take place. The transfer of only a fraction of the amount of money which would have been transferred under a purely metallic system, and the substitution of a multiple credit contraction for the rest, as it were, deprives the individuals in the country concerned of the possibility of delaying the adaptation by temporarily paying for an excess of imports in cash.

That the rise of the rate of interest in the country that

[1] This has been rightly pointed out, but has hardly been sufficiently explained, in an interesting article by J. C. Gilbert on the Present Position of the Theory of International Trade, *The Review of Economic Studies*, vol. III, no. 1, October 1935, particularly pp. 23-6 — To say that money rates of interest in a particular country may be made to deviate from the equilibrium rate by monetary factors peculiar to that country is of course not to say that the equilibrium rate in that country is independant of international conditions.

is losing gold, and the corresponding reduction in the bank rate in the country which is receiving gold, need have nothing to do with changes in the demand for or the supply of capital appears also from the fact that, if no further change intervenes, the new rates will have to be kept in force only for a comparatively short period, and that after a while a return to the old rates will be possible. The changes in the rates serve the temporary purpose of speeding up a process which is already under way. But the forces which would have brought the flow of gold to an end earlier or later in any case do not therefore cease to operate. The chain of successive reductions of income in country A set up by the initiating changes will continue to operate and ultimately reduce the payments out of the country still further. But since payments in and payments out have in the meantime already been made to balance by the action of the banks, this will actually reverse the flow and bring about a favourable balance of payments. The banks, wanting to replenish their reserves, may let this go on for a while, but once they have restored their reserves, they will be able to resume at least the greater part of their lending activity which they had to curtail.

This picture is admittedly incomplete because I have been deliberately neglecting the part played by short term capital movements. I shall discuss these in my fourth lecture. At present my task merely is to show how the existence of national banking systems, based on the collective holding of national cash reserves, alters the effects of international flows of money. It seems to me impossible to doubt that there is indeed a very considerable difference between the case where a country, whose inhabitants are induced to decrease their share in the world's stock of money by ten per cent, does so by actually giving up this ten per cent in gold, and the case where, in order to preserve the accustomed reserve proportions, it pays out only one per cent in gold and contracts the credit super-structure in proportion to the reduction of reserves. It is as if all balances of international payments had to be squeezed through a narrow bottle neck, as special pressure has to be brought on people who would otherwise not have been affected by the change, to give up money which they would have invested productively.

Now the changes in productive activity which are made necessary in this way are not of a permanent nature. They mean not only that in the first instance many plans will be upset, that equipment which has been created will cease to be useful and that people will be thrown out of employment. They also mean that the revised plans which will be made are bound soon to be equally disappointed in the reverse direction and that the readjustment of production which has been enforced will prove to be a misdirection. In other words, it is a disturbance which possesses all the characteristics of a purely monetary disturbance, namely that it is self-reversing in the sense that it induces changes which will have to be reversed because they are not based on any corresponding change in the underlying real facts.

It might perhaps be argued that the contraction of credit in the one country and the expansion in the other brings about exactly the same effects that we should expect from a transfer of a corresponding amount of capital from the one country to the other, and that since the amount of money which would otherwise have to be transferred would represent so much capital, there can be no harm in the changes in the credit structure. But the point is exactly that not every movement of money is in this sense a transfer of capital. If a group of people want to hold more money because the value of their income rises, while another group of people reduce their money holdings because the value of their income falls, there is no reason why in consequence the funds available for investment in the first group should increase and those available in the second group should decrease. It is, on the other hand, quite possible that the demand for such funds in the first group will rise and in the second group will fall. In such a case, as we have seen, there would be more reason to expect that the rate of interest will rise in the country to which the money flows rather than in the country from which the money comes.

The case is of course different when the initiating cause is not a shift in demand from one kind of consumers' goods to another kind of consumers' goods, but when funds which have been invested in one type of producers' goods in one country are transferred to investment in another type of producers' goods in another country Then indeed we have a true movement of capital and we should be entitled to expect it to affect interest rates in the usual manner. What I am insisting on is merely that this need not be the general rule and that the fact that it is generally the case is not the effect of an inherent necessity but due to purely institutional reasons.

4

There are one or two further points which I must shortly mention before I can conclude this subject. One is the rather obvious point that the disturbing effects of the organisation of the world's monetary system on the national reserve principle are of course considerably increased when the rate of multiple expansion or contraction, which will be caused by a given increase or decrease of gold, is different in different countries. If this is the case, and it has of course always been the case

INTERNATIONAL FLOWS OF MONEY 33

under the gold standard as we knew it, it means that every flow of gold from one country to another will mean either an inflation or a deflation from the world point of view, accordingly as the rate of secondary expansion is greater or smaller in the country receiving gold than in the country losing gold.

The second point is one on which I am particularly anxious not to be misunderstood. The defects of the mixed system which I have pointed out are not defects of a particular kind of policy, or of special rules of central bank practice. They are defects inherent in the system of the collective holding of proportional cash reserves for national areas, whatever the policy adopted by the central bank or the banking system. What I have said provides in particular no justification for the common infringements of the " rules of the game of the gold standard ", except, perhaps for a certain reluctance to change the discount rate too frequently or too rapidly when gold movements set in. But all the attempts to substitute other measures for changes in the discount rate as a means to "protect reserves " do not help, because it is the necessity of " protecting " reserves rather than letting them go (*i.e.* using the conversion into gold as the proper method of reducing internal circulation) not the methods by which it has to be done, which is the evil. The only real cure would be if the reserves kept were large enough to allow them to vary by the full amount by which the total circulation of the country might possibly change; that is, if the principle of Peel's Act of 1844 could be applied to all forms of money, including in particular bank deposits. I shall come back to this point in my last lecture. What I want to stress, however, is that in the years before the breakdown of the international gold standard the attempts to make the supply of money of individual countries independent of international gold movements had already gone so far that not only had an outflow or inflow of gold often no effect on the internal circulation but that sometimes the latter moved actually in the opposite direction. To " offset " gold movements, as was apparently done by the Bank of England,[1] by replacing the gold lost by the central bank by securities bought from the market, is of course not to correct the defects of the mixed system, but to make the international standard altogether ineffective.

One should probably say much more on this subject. But I am afraid I must conclude here. I hope that what I have said to-day has at least made one point clear which I made yesterday; namely that many objections which are raised against the gold standard as we knew it, are not really objections against the gold standard, or against any international standard as such, but objections against the mixed system which has been in general vogue. It should be clear too that the main defect of this system was that it was not sufficiently international. Whether and how these defects can be remedied I can consider only at the end of this course. But before I can do this I shall yet have to consider the more completely nationalist systems which have been proposed.

---

[1] Cf. *Minutes of Evidence taken before the Committee on Financ and Industry*, London, 1931, vol. I, Q. 353. Sir Ernest Harvey " You will find if you look at a succession of Bank Returns tha the amount of gold we have lost has been almost entirely replace by an increase in the Bank's securities. "

LECTURE III

INDEPENDENT CURRENCIES

1

When the rates of exchange between currencies of different countries are variable, the consequences which will follow from changes which under an international system would lead to flows of money from country to country, will depend on the monetary policies adopted by the countries concerned. It is therefore necessary, before we can say anything about those effects, to consider the aims which will presumably guide the monetary policy of countries which have adopted an independent standard. This raises immediately the question whether there is any justification for applying any one of the principles according to which we might think that the circulation in a closed system should be regulated, to a particular country or region which is part of the world economic system.

Now it should be evident that a policy of stabilization, whether it be of the general price level or the general level of money incomes is one thing if it be applied to the whole of a closed system and quite another if the same policy is applied to each of the separate regions into which the total system can be more or less arbitrarily divided. In fact, however, this difficulty is generally ignored by the advocates of Monetary Nationalism, and it is simply assumed that the criteria of a good monetary policy which are applicable to a closed system are equally valid for a single country. We shall have to consider later the theoretical problems here involved. But for the moment we can confine ourselves to an examination of the working of the mechanism which brings about relative changes in the value of the total money holdings of the different nations, when each nation follows independently the objective of stabilising its national price level, or income stream, or whatever it may be, irrespective of its position in the international system.

The case which has figured most prominently in these discussions in recent years, and which is apparently supposed to represent the relative positions of England and the United States, is that of two countries with unequal rates of technological progress, so that, in the one, costs of production will tend to fall more rapidly than in the other. Under a regime of fixed parities this would mean that the fall in the prices of some products produced in both countries could be faster than the fall in their cost in the country where technological progress is slower, and that in consequence it would become necessary to reduce costs there by scaling down money wages, etc. The main advantage of a system of movable parities is supposed to be that in such a case the downward adjustment of wages could be avoided and equilibrium restored by reducing the value of money in the one country relative to the other country.

It is, however, particularly important in this connection not to be misled by the fact that this argument is generally expressed in terms of averages, that is in terms of general levels of prices and wages. A change in the level of prices or of costs in one country relatively to that of another means that, in consequence of changes in relative costs, the competitive position of a particular industry or perhaps group of industries in the one country has deteriorated. In other words the lower prices in the one country will lead to a transfer of demand from the other country to it. The case is therefore essentially similar to that which we have been considering in the last lecture and it will be useful to discuss it in the same terms. We shall therefore in the first instance again consider the effects of a simple shift of demand if rates of exchange are allowed to vary and if the monetary authorities in each country aim either at stability of some national price level, or—what amounts very much to the same thing for our purpose—at a constant volume of the effective money stream within the country. Only occasionally, where significant differences arise, I shall specially refer to the case where the shift of demand has been induced by unequal technological progress.

Now of course no monetary policy can prevent the prices of the product immediately affected from falling relatively to the prices of other goods in the one country, and a corresponding[1] rise taking place in the other. Nor can it prevent the effects of the change of the income of the people affected in the first instance from gradually spreading. All it can do is to prevent this from leading to a change in the *total* money stream in the country; that is it must see that there will be offsetting changes of other prices so that the price level remains constant. It is on this assumption that we conduct our investigations. For purposes of simplicity, too, I assume that at the outset a state of full employment prevails.

---

[1] Where the shift of demand has been induced by a reduction of cost and a consequent fall of prices in the one country, this will only be a relative rise and will of course only partly counteract this fall in the price of the final product, but may bring about an actual rise in the prices of the factors used in their production.

2

It will be convenient to concentrate first on the country from which demand has turned away and from which under an international monetary system there would in consequence occur an outflow of money. But in the present case not only would a real outflow of money be impossible, but it would also be contrary to the intentions of the monetary authorities to sell additional quantities of foreign exchange against national money and to cancel the national money so received. The monetary authorities might hold some reserves of foreign exchange to even out what they regarded as merely temporary fluctuations of exchange rates. But there would be no point in using them in the case of a change which they would have to regard as permanent. We can, therefore, overlook the existence of such reserves and proceed as if only current receipts from abroad were available for outward payments.

On this assumption it is clear that the immediate effect of the adverse balance of payments will be that the foreign exchanges will rise. But the full amount that importers used to spend on buying foreign exchange is not likely to be spent on the reduced supply of foreign exchange; since with the higher price of imported goods some of the money which used to be spent on them will probably be diverted to home substitutes.[1] The foreign exchanges will therefore probably rise less than in proportion to the fall in supply. But *via* the sale of foreign exchange at

---

[1] The assumption that the demand for the commodities in question is elastic, that is that the total expenditure upon them will be reduced when their prices rise and *vice versa*, will be maintained throughout this discussion. To take at every step the opposite case into account would unduly lengthen the argument without affecting the conclusion.

INDEPENDENT CURRENCIES 39

the higher rate those who continue to export successfully will receive greater amounts of the national currency. For those whose sales abroad have not been unfavourably affected by the initial change in question this will mean a net gain and the price of their products will correspondingly rise in terms of the national currency. And those whose exports have fallen in price will find that this reduced price in terms of the foreign currency will now correspond to a somewhat greater amount in the national currency than what they could obtain before the exchange depreciation, although not as much as they received before the first change took place.

This impact effect of the rise of exchange rates on relative prices in terms of the national currency will however be temporary. The relative costs of the different quantities of the different commodities which are being produced have not changed and it is not likely that they will go on being produced in these quantities if their prices have changed. Moreover all the changes in the direction of the money streams caused by the rise in exchange rates will continue to work. More is being spent on home goods, and this, together with the increased profitability of those export industries which have not been adversely affected by the initial change, will tend to bring about a rise of all prices except those which are affected by the decreased demand from the declining industry and from the people who draw their income from it.

It seems therefore that the argument in favour of depreciation in such cases is based on a too simplified picture of the working of the price mechanism. In particular it seems to be based on the assumption (underlying much of the classical analysis of these problems) that relative prices within each country are uniquely determined by (constant) relative cost. If this were so, a proportional reduction of all prices in a country relatively to those in the rest of the world would indeed be sufficient to restore equilibrium. In fact, however, there can be little doubt that the changes in the relative quantities of goods to be produced by the different industries which will become necessary in consequence of the initial change, can be brought about only by changes in the relative prices and the relative incomes of the different kinds of resources within the country.

Without following the effects in all their complicated detail it must be clear that the ultimate result of depreciation can only be that, instead of prices and incomes in the industry originally affected falling to the full extent, a great many other prices and incomes will have to rise to restore the proportions appropriate to cost conditions and the relative volume of output now required. Even disregarding the absolute height of prices, the final positions will not be the same as that which would have beeen reached if exchanges had been kept fixed; because in the course of the different process of transition all sorts of individual profits and losses will have been made which will affect that final position. But roughly speaking and disregarding certain minor differences, it can be said that the same change in relative prices which, under fixed exchanges, would have been brought about by a reduction of prices in the industry immediately affected is now being brought about largely by a corresponding rise of all other prices.

Two points, however, need special mention. One is that the decrease of the comparative advantage of the export industry originally affected cannot be changed in this way; and that to this extent a contraction of the output of this industry will remain unavoidable. The other is that at least in certain respects the process which brings about the rise in prices will be of a definitely inflationary character. This will show itself partly by some

INDEPENDENT CURRENCIES 41

industries becoming *temporarily* more profitable so that there will be an inducement to expand production there, although this increase will soon be checked and even reversed by a rise in cost; and partly by some of the cash released by importers finding its way, *via* the repayment of loans, to the banks, who will be able to increase their loans to others and, in order to find lenders, will relax the terms on which they will be ready to lend. But this too will prove a merely temporary effect, since as soon as costs begin generally to rise it will become apparent that there are really no funds available to finance additional investments. In this sense the effects of this redistribution of money will be of that self-reversing character which is typical of monetary disturbances. This leads, however, already to the difficult question of what constitutes an inflation or deflation within a national area. But before we can go on to this it is necessary to consider what happens in the converse case of the country which has been put in a more favourable condition by the change.

3

Let us first assume that the monetary authorities here as in the other country aim at a constant price level and a constant income stream. The industry which directly benefits from the initial shifts in demand will then find that, because of the fall of foreign exchanges, the increase of their receipts in terms of the national currency will not be as large as would correspond to the increase of their sales in terms of foreign money, while the other export industries will see their receipts actually reduced. Similarly those home industries whose products compete with imports which are now cheaper in terms of the national currency will have to lower their prices and will find their incomes reduced. In short, if the quantity of money in the country, or the price level, is kept constant, the increase of the aggregate value of the products of one industry due to a change in international demand will mean that there has to be a compensating reduction of the prices of the products of other industries. Or, in other words, part of the price reduction which under a regime of stable exchanges would have been necessary in the industry and in the country from which demand has turned away, will under a regime of independent currencies and national stabilisation have to take place in the country towards which demand has turned, and in industries which have not been directly affected by the shift in demand.

This at least should be the case if the principle of national stabilization were consistently applied. But it is of course highly unlikely that it ever would be so applied. That in order to counteract the effects of a severe fall of prices in one industry in a country other prices in the country should be allowed to rise, appears fairly plausible. But that in order to offset a rise of prices of the products of one industry which is due to an increase in international demand, prices in the other industries should be made to fall sounds far less convincing. I find it difficult to imagine the President of a Central Bank explaining that he has to pursue a policy which means that the prices of many home industries have to be reduced, by pointing out that an increase of international demand has led to an increase of prices in an important export industry, and it seems fairly certain what would happen to him if he tried to do so.

Indeed, if we take a somewhat more realistic point of view, there can be little doubt what will happen. While, in the country where in consequence of the changes in international demand some prices will tend to fall the price level will be kept stable, it will certainly be allowed to rise in the country which has been benefited by the same shift in demand. It is not difficult to see what this implies if all countries in the world act on this principle. It means that prices would be stabilized only in that area where they tend to fall lowest relatively to the rest of the world, and that all further adjustments are brought about by proportionate increases of prices in all other countries. The possibilities of inflation which this offers if the world is split up into a sufficient number of very small separate currency areas seem indeed very considerable. And why, if this principle is once adopted, should it remain confined to average prices in particular national areas? Would it not be equally justified to argue that no price of any single commodity should ever be allowed to fall and that the quantity of money in the world should be so regulated that the price of that commodity which tends to fall lowest relatively to all others should be kept stable, and that the prices of all other commodities would be adjusted upwards in proportion? We only need to remember what happened, for instance, a few years ago to the price of rubber to see how such a policy would surpass the wishes of even the wildest inflationist. Perhaps this may be thought an extreme case. But, once the principle has been adopted, it is difficult to see how it could be confined to " reasonable " limits, or indeed to say what " reasonable " limits are.

4

But let us disregard the practical improbability that a policy of stabilization will be followed in the countries where, with stable exchanges, the price level would rise, as well as in the countries where in this case it would have to fall. Let us assume that, in the countries which benefit from the increase of the demand, the prices of other goods are actually lowered to preserve stability of the national price level and that the opposite action will be taken in the countries from which demand has turned away. What is the justification and significance of such a policy of national stabilization?

Now it is difficult to find the theoretical case for national stabilization anywhere explicitly argued. It is usually just taken for granted that any sort of policy which appears desirable in a closed system must be equally beneficial if applied to a national area. It may therefore be desirable before we go on to examine its analytical justification, to trace the historical causes which have brought this view to prominence. There can be little doubt that its ascendancy is closely connected with the peculiar difficulties of English monetary policy between 1025 and 1931. In the comparatively short space of the six years during which Great Britain was on a gold standard in the post war period, it suffered from what is known as overvaluation of the pound. Against all the teaching of " orthodox " economics—already a hundred years before Ricardo had expressly stated that he " should never advise a government to restore a currency, which was depreciated 30 p. c., to par "[1]—in 1925 the British currency had been brought back to its former gold value. In consequence, to restore equilibrium, it was necessary to reduce *all* prices and costs in proportion as the value of the pound had been raised. This process, particularly because of the notorious difficulty of reducing money wages, proved to be very painful and prolonged. It deprived England of real participation in the boom which led up to the crisis of 1929, and, in the end, its results

[1] In a letter to John Wheatley, dated September 18, 1821, reprinted in *Letters of David Ricardo to Hutches Trower and Others*, edited by J. Bonar and J. Hollander, Oxford, 1899, p. 160.

INDEPENDENT CURRENCIES 45

proved insufficient to secure the maintenance of the restored parity. But all this was not due to an initial shift in the conditions of demand or to any of the causes which may affect the condition of a particular country under stable exchanges. It was an *effect* of the change in the external value of the pound. It was not a case where with given exchange rates the national price or cost structure of a country as a whole had got out of equilibrium with the rest of the world, but rather that the change in the parities had suddenly upset the relations between all prices inside and outside the country.

Nevertheless this experience has created among many British economists a curious prepossession with the relations between national price- and cost- particularly wage-levels, as if there were any reason to expect that as a rule there would arise a necessity that the price and cost structure of one country as a whole should change relatively to that of other countries. And this tendency has received considerable support from the fashionable pseudo-quantitative economics of averages with its argument running in terms of national " price levels ", " purchasing power parities ", " terms of trade ", the " Multiplier ", and what not.

The purely accidental fact that these averages are generally computed for prices in a national area is regarded as evidence that in some sense all prices of a country could be said to move together relatively to prices in other countries.[1] This has strengthened the belief that there

---

[1] The fact that the averages of (more or less arbitrarily selected) groups of prices move differently in different countries does of course in no way prove that there is any tendency of the price structure of a country to move as a whole relatively to prices in other countries. It would however be a highly interesting subject for statistical investigation, if a suitable technique could be devised, to see whether, and to what extent, such a tendency existed. Such an investigation would of course involve a comparison

is some peculiar difficulty about the case where " the " price level of a country had to be changed relatively to its given cost level and that such adjustment had better be avoided by manipulations of the rate of exchange.

Now let me add immediately that of course I do not want to deny that there may be cases where some change in conditions might make fairly extensive reductions of money wages necessary in a particular area if exchange rates are to be maintained, and that under present conditions such wage reductions are at best a very painful and long drawn out process. At any rate in the case of countries whose exports consist largely of one or a few raw materials, a severe fall in the prices of these products might create such a situation. What I want to suggest, however, is that many of my English colleagues, because of the special experience of their country in recent times, have got the practical significance of this particular case altogether out of perspective : that they are mistaken in believing that by altering parities they can overcome many of the chief difficulties created by the rigidity of wages and, in particular, that by their fascination with the relation between " the " price level and " the " cost level in a particular area they are apt to overlook the much more important consequences of inflation and deflation.[1]

---

not only of some mean value of the price changes in different countries, but of the whole frequency distribution of relative price changes in terms of some common standard. And it should be supplemented by similar investigations of the relative movements of the price structure of different parts of the same country.

[1] The propensity of economists in the Anglo-Saxon countries to argue exclusively in terms of national price and wage levels is probably mainly due to the great influence which the writings of Professor Irving Fisher have exercised in these countries. Another typical instance of the dangers of this approach is the well-known controversy about the reparations problem, where it was left to Professor Ohlin to point out against his English opponents that

5

As I have already suggested at an earlier point, the difference of opinion here rests largely on a difference of view on the meaning and consequence of inflation and deflation, or rather in the importance attached to two sorts of effects which spring from changes in the quantity of money. The one view stresses what I have called before the self-reversing character of the effects of monetary changes. It emphasizes the misdirection of production caused by the wrong expectation created by changes in relative prices which are necessarily only temporary, of which the most conspicuous is of course the trade cycle. The other view emphasizes the effects which are due to the rigidity of certain money prices, and particularly wages. Now the difficulties which arise when money wages have to be lowered can not really be called monetary disturbances; the same difficulties would arise if wages were fixed in terms of some commodity. It is only a monetary problem in the sense that this difficulty might to some extent be overcome by monetary means when wages are fixed in terms of money. But the problem left unanswered by the authors who stress this second aspect is whether the difficulty created by the rigidity of money wages can be overcome by monetary adjustments without setting up new disturbances of the first kind. And there are in fact strong reasons to believe that the two aims of avoiding so far as possible downward adjustments of wages and preventing misdirections of production may not always be reconcilable.

This difference in emphasis is so important in connection with the opinions about what are the appropriate

what mainly mattered was not so much effects on total price levels but rather the effects on the position of particular industries.

principles of national monetary policy, because if one thinks principally in terms of the relation of prices to given wages, particularly if one thinks in terms of national wage " levels ", one is easily led to the conclusion that the quantity of money should be adjusted for each group of people among whom a given system of contracts exists. (To be consistent, of course, the argument should be applied not only to countries but also to particular industries, or at any rate to " non-competing groups " of workers in each country.) On the other hand, there is no reason why one should expect the self-reversing effects of monetary changes to be connected with the change of the quantity of money in a particular area which is part of a wider monetary system. If a decrease or increase of demand in one area is offset by a corresponding change in demand in another area, there is no reason why the changes in the quantity of money in the two areas should in any sense misguide productive activity. They are simply manifestations of an underlying real change which works itself out through the medium of money.

To illustrate this difference let me take a statement of one of the most ardent advocates of Monetary Nationalism, Mr. R. F. Harrod of Oxford. Mr. Harrod is not unfamiliar with what I have called the self-reversing effects of monetary changes. At any rate in an earlier publication he argued that " if industry is stimulated to go forward at a pace which cannot be maintained, you are bound to have periodic crises and depressions "[1]. Yet

---

[1] *The International Gold Problem*, Edited by the Royal Institute of International Affairs. London, 1931, p. 29. Cf. also, in the light of this statement, the remarkable passage in the same author's *International Economics (London*, 1933), p. 150, where it is argued, that " the only way to avoid a slump is to engineer a boom " although only two lines later a boom is still " defined as an increase in the rate of output which cannot be maintained in the long period ".

INDEPENDENT CURRENCIES 49

for some reason he seems to think that these misdirections of industry will occur even when the changes in the quantity of money of a particular country take place in the course of the normal redistributions of money between countries. In his *International Economics* there appears the following remarkable passage which seems to express the theoretical basis, or as I think the fallacy, underlying Monetary Nationalism more clearly than any other statement I have yet come across. Mr. Harrod is discussing the case of unequal economic progress in different countries with a common standard and concludes that " the less progressive countries would thus be afflicted with the additional inconvenience of a deflatory monetary system. Inflation would occur just where it is most dangerous, namely in the rapidly advancing countries. This objection appears in one form or another in all projects for a common world money ".[1] And the lesson which Mr. Harrod derives from these considerations is that " the currencies of the more progressive countries must be made to appreciate in terms of the others ".[2]

It is interesting to inquire in what sense inflation and deflation are here represented as *additional* inconveniences, superimposed, as it were, on the difficulties created by unequal economic progress. One might think at first that what Mr. Harrod has in mind are the *extra* difficulties caused by the secondary expansions and contractions of credit which are made necessary by the national reserve systems which I have analyzed in an earlier lecture. But this interpretation is excluded by the express assertion that this difficulty appears under *all* forms of a common world money. It seems that the terms inflation and deflation are here used simply as equivalents to

[1] *International Economics* (Cambridge Economic Handbooks VIII), London, 1933, p. 170.
[2] *Ibid.*, p. 174.

increases and decreases of money demand relatively to given costs. In this sense the terms could equally be applied to shifts in demand between different industries and would really mean no more than a change in demand relatively to supply. But the objection to this is not only that the terms inflation and deflation are here unnecessarily applied to phenomena which can be described in simpler terms. It is rather whether in this case there is any reason to expect any of the special consequences which we associate with monetary disturbances, that is, whether there really is any " additional inconvenience " caused by monetary factors proper. We might ask whether in this case there will be any of the peculiar self-reversing effects which are typical of purely monetary causes; in particular whether " inflation " as used here with reference to the increase of money in one country at the expense of another, " stimulates industry to go forward at a pace which cannot be maintained "; and whether deflation in the same sense implies a temporary and avoidable contraction of production.

The answer to these questions is not difficult. We know that the really harmful effects of inflation and deflation spring, not so much from the fact that all prices change in the same direction and in the same proportion, but from the fact that the relation between individual prices changes in a direction which cannot be maintained; or in other words that it temporarily brings about a distribution of spending power between individuals which is not stable. We have seen that the international redistributions of money are part of a process which at the same time brings about a redistribution of relative amounts of money held by the different individuals in each country, a redistribution within the nation which would also have to come about if there were no international money. The difference, however, in the latter

INDEPENDENT CURRENCIES 51

case, the case of free currencies, is that here first the relative value of the total amounts of money in each country is changed and that the process of internal redistribution takes places in a manner different from that which would occur with an international monetary standard. We have seen before that the variation of exchange rates will in itself bring about a redistribution of spending power in the country, but a redistribution which is in no way based on a corresponding change in the underlying real position. There will be a temporary stimulus to particular industries to expand, although there are no grounds which would make a lasting increase in output possible. In short, the successive changes in individual expenditure and the corresponding changes of particular prices will not occur in an order which will direct industry from the old to the new equilibrium position. Or, in other words, the effects of keeping the quantity of money in a region or country constant when under an international monetary system it would decrease are essentially inflationary, while to keep it constant if under an international system it would increase at the expense of other countries would have effects similar to an absolute deflation.

I do not want to suggest that the practical importance of the deflationary or inflationary effects of a policy of keeping the quantity of money in a particular area constant is very great. The practical arguments which to me seem to condemn such a policy I have already discussed. The reason why I wanted at least to mention this more abstract consideration is that, if it is correct, it shows particularly clearly the weakness of the theoretical basis of Monetary Nationalism. The proposition that the effects of keeping the quantity of money constant in a territory where with an international currency it would decrease are inflationary and *vice-versa*[1] is of course directly contrary to the position on which Monetary Nationalism is based. Far from admitting that changes in the relative money holdings of different nations which go parallel with changes in their share of the world's income are harmful, we believe that such redistributions of money are the only way of effecting the change in real income with a minimum of disturbance. And to speak in connection with such changes of national inflation or deflation can only lead to a serious confusion of thought.[2]

Before I leave this subject I should like to supplement these theoretical reflections by a somewhat more practical consideration. While the whole idea of a monetary policy directed to adjust everything to a " given " wage level appears to me misconceived on purely theoretical grounds, its consequences seem to me to be fantastic if we imagine it applied to the present world where this supposedly given wage level is at the same time the subject

---

[1] Whithout giving disproportionate space to what is perhaps a somewhat esoteric theoretical point it is not possible to give here a complete proof of this proposition. A full discussion of the complicated effects would almost require a separate chapter. But a sort of indirect proof may be here suggested. It would probably not be denied that if without any other change the amount of money in one currency area were decreased by a given amount and at the same time the amount of money in another currency area increased by a corresponding amount, this would have deflationary effects in the first area and inflationary effects in the second. And most economists (the more extreme monetary nationalists only excepted) would agree that no such effects would occur if these changes were made simultaneous with corresponding changes in the relative volume of transactions in the two countries. From this it appears to follow that if such a change in the relative volume of transactions in the two countries occurs but the quantity of money in each country is kept constant, this must have the effect of a relative inflation and deflation respectively.

[2] See on this point also L. ROBBINS, *Economic Planning and International Order*, 1937, pp. 281 et seq.

INDEPENDENT CURRENCIES 53

of political strife. It would mean that the whole mechanism of collective wage bargaining would in the future be used exclusively to raise wages, while any reduction— even if it were necessary only in one particular industry—would have to be brought about by monetary means. I doubt whether such a proposal could ever have been seriously entertained except in a country and in a period where labour has been for long on the defensive.[1] It is difficult to imagine how wage negotiations would be carried on if it became the recognised duty of the monetary authority to offset any unfavourable effect of a rise in wages on the competitive position of national industries on the world market. But of one thing we can probably be pretty certain : that the working class would not be slow to learn that an engineered rise of prices is no less a reduction of wages than a deliberate cut of money wages, and that in consequence the belief that it is easier to reduce by the round-about method of depreciation the wages of all workers in a country than directly to reduce the money wages of those who are affected by a given change, will soon prove illusory.

---

[1] It is interesting to note that those countries in Europe where up to 1929 wages had been rising relatively most rapidly were on the whole those most reluctant to experiment with exchange depreciation. The recent experience of France seems also to suggest that a working class government may never be able to use exchange depreciation as an instrument to lower real wages.

LECTURE IV

INTERNATIONAL CAPITAL MOVEMENTS

1

For the purposes of this lecture, by international capital movements I shall mean the acquisition of claims on persons or of rights to property in one country by persons in another country, or the disposal of such claims or property rights in another country to people in that country. This definition is meant to exclude from capital movements the purchase and sale of commodities which pass from one country to the other at the same time as they are paid for and change their owners. But it also excludes any net movement of gold (or other international money) in so far as these movements are payments for commodities or services received (or " unilateral " payments) and therefore involve a transfer of ownership in that money without creating a new claim from one country to the other. This is of course not the only possible definition of capital movements, and strong arguments could be advanced in favour of a more comprehensive definition, which in effect would treat every transfer of assets from country to country as a capital movement. The reason which leads me to adopt here the former definition is that only on that definition it is possible to distinguish between those items in international transactions which are, and those which are not, capital items.

The first kind of capital item of this sort and the one which will occupy us in this lecture more than any other is the acquisition, or sale, of amounts of the national

INTERNATIONAL CAPITAL MOVEMENTS 55

money of one country by inhabitants of the other.[1] The form which this kind of transaction to-day predominantly takes is the holding of balances with the banks of one country on the part of banks and individuals in the other country. Such balances will to some extent be held even if there is a safe and stable international standard, since, rather than actually send money, it will as a rule be cheaper for the banks to provide out of such balances those of their customer's requirements which arise out of the normal day to day differences between payments and receipts abroad. And if it is possible to hold such balances either in the form of interest-bearing deposits or in the form of bills of exchange, there will be a strong inducement to hold such earning assets as substitutes for the sterile holdings of international money. It was in this way that what is called the gold exchange standard tended more and more to supplant the gold standard proper. In the years immediately preceding 1931 this assumed very great significance.

If there exists a system of fluctuating exchanges, or a system where people are not altogether certain about the maintenance of the existing parities, these balances become even more important. There are two new elements which enter in this case. In the first place it will here no longer be sufficient if banks and others who owe debts in different currencies keep one single liquidity reserve against all their liabilities. It will become neces-

---

[1] This is not to be interpreted as meaning that I subscribe to the view that all money is in some sense a " claim ". The statement in the text applies strictly only to credit money and particularly to bank deposits, which will be mainly considered in what follows. But it would not apply to the acquisition of gold by foreigners for export. The gold coins so acquired would thereby cease to be " national " money in the sense in which this term is here used, that is, they would not be assets belonging to the country where they have been issued.

sary for them to keep separate liquid assets in each of the different currencies in which they owe debts, and to adjust them to the special circumstances likely to affect liabilities in each currency. We get here new artificial distinctions of liquidity created by the multiplicity of currencies and involving all the consequential possibilities of disturbances following from changes in what is now called "liquidity preference". Secondly there will be the chance of a gain or loss on these foreign balances due to changes in the rates of exchange. Thus the anticipation of any impending variation of exchange rates will tend to bring about temporary changes of a speculative nature in the volume of such balances. Whether these two kinds of motives must really be regarded as different, or whether they are better treated as essentially the same, there can be no doubt that variability of exchange rates introduces a new and powerful reason for short term capital movements, and a reason which is fundamentally different from the reasons which exist under a well-secured international standard.

Foreign bank balances and other holdings of foreign money are of course only part, although probably the most important part, of the volume of short term foreign investment. It is here that the impact effect of any change in international indebtedness arising out of current transactions will show itself; and it is here that there will be the most ready response to changes in the relative attractiveness of holding assets in the different countries. Once we go beyond this field it becomes rather difficult to say what can properly be called movements of short term capital. In fact, with the exception of non-funded long-term loans almost any form of international investment may have to be regarded as short term investment, including in particular all investments in marketable securities.[1] But for the monetary problems with which we are here concerned it is mainly the short term credits which are of importance, because it is here that we have to deal with large accumulated funds which are apt to change their location at comparatively slight provocation. Compared with these " floating " funds, the supply of capital for long term investment, limited as it will be to a certain part of new savings, will be relatively small.

Now the chief question which we shall have to consider is the question to what extent under different monetary systems international capital movements are likely to cause monetary disturbances, and to what extent and by what means it may be possible to prevent such disturbances. It will again prove useful if we approach this task in three stages, beginning with a consideration of the mechanism and function of international capital movements under a homogeneous standard. Then we can go on to inquire how this mechanism and the effects are modified if we have " mixed " currency systems organized on the national reserve principle but with fixed exchange rates. And finally we shall have to see what will be the effects of the existence of variable exchange rates and the way in which fluctuations of the exchange and capital movements mutually influence one another.

2

If exchange rates were regarded as invariably fixed we should expect capital movements to be guided by no other

[1] Even the intentions of the lender or investor would hardly provide a sufficient criterion for a distinction between what are short and what are long term capital movements, since it may very well be clear in a particular case to the outside observer that circumstances will soon lead the investors to change their intentions.

considerations except expected net yield, including of course adjustments which will have to be made for the different degrees of risk inherent in the different sorts of investments. This does not mean that there would not be frequent changes in the flow of capital from country to country. There might of course be a permanent tendency on the part of one country to absorb part of the current savings of another at terms more favourable than those at which these savings could be invested in the country were they are made. Quite apart from these flows of capital for more or less permanent investment however, there would be periodic or occasional short term lending to make up for temporary differences between imports and exports of commodities and services.

Now there is of course no reason why exports and imports should move closely parallel from day to day or even from month to month. If in all transactions payment had to be made simultaneously with the delivery of the goods, this would mean, in external trade no less than in internal, a restriction of the possible range of transactions similar in kind to what would occur if all transaction had to take the form of barter. The possibility of credit transactions, the exchange of present goods against future goods, greatly widens the range of advantageous exchanges. In international trade it means in particular that countries may import more than they export in some seasons because they will export more than they import during other seasons. Whether this is made possible by the exporter directly crediting the importer with the price, or whether it takes place by some credit institution in either country providing the money, it will always mean that the indebtedness of the importing country to the exporting increases temporarily, that is, that net short term lending takes place.

At this point it is necessary especially to be on guard

INTERNATIONAL CAPITAL MOVEMENTS 59

against a form of stating these relations which suggests that short term lending is made necessary by, or is in any sense a consequence of a passive balance of trade; that the loans are made so to speak with the purpose of covering a deficit in the balance of trade. We shall get a more correct picture if we think of the great majority of the individual transactions in both ways being credit transactions so that it is the excess lending in one direction during any given period which has made possible a corresponding excess of exports in the same direction. If we look on the whole process in this way we can see how considerable a part of trade is only made possible by short term capital movements. We can see also how misleading it may be to think of capital movements as exclusively directed by previous changes in the relative rates of interest in the different money markets. What directs the use of the available credit and therefore decides in what direction the balance of indebtedness will shift at a particular moment is in the first instance the relation between prices in different places. It is of course true that where each country habitually finances its exports and borrows its imports, any absolute increase of exports will tend to bring about an increase in the demand for loans and therefore a rise in the rate of interest in the exporting country. But in such a case the rise in the rate of interest is rather the effect of this country lending more abroad, than a cause of a flow of capital to the country. And although this rise in money rates may lead to a flow of funds in the reverse direction, that will be more a sign that the main mechanism for the distribution of funds works imperfectly than a part of this mechanism. There is no more reason to say that the international redistribution of short-term capital is brought about by changes in the rates of interest in the different localities than there would be for saying that the seasonal transfers of funds from say agriculture to coal mining are brought about by a fall of the rate of interest in agriculture and a rise in coal mining and *vice versa*.

Changes in short term international indebtedness must therefore be considered as proceeding largely concurrently with normal fluctuations in international trade; and only certain remaining balances will be settled by a flow of funds, largely of an inter-bank character, induced by differences in interest rates to be earned. It is of course not to be denied that, apart from changes in international indebtedness which are more directly connected with international trade, there may also be somewhat sudden and considerable flows of funds which may be caused either by the sudden appearance of very profitable opportunities for investment, or by some panic which causes there an insistent demand for cash. In this last case indeed it is true that the flow of short term funds may transmit monetary disturbances to parts of the world which have nothing to do with the original cause of the disturbance, as say a war-scare in South-America might conceivably lead to a general rise in interest rates in London. But, apart from such special cases, it is difficult to see how under a homogeneous international standard, capital movements, and particularly short term capital movements, should be a source of instability or lead to any changes in productive activity which are not justified by corresponding changes in the real conditions.

3

This conclusion has, however, to be somewhat modified if, instead of a homogeneous international currency, we consider a world consisting of separate national monetary and banking systems, even if we still leave the possibility of variations in exchange rates out of account.

It is of course a well-known fact, that one of the main purposes of changes in the discount rate of central banks is to influence the international movements of short term capital.[1] A central bank which is faced with an outflow of gold will raise its discount rate in the hope that by attracting short term credits it will offset the gold outflow. To the extent that it succeeds it will postpone the necessity of more drastic credit contraction at home, and—if the cause of the adverse balance of trade is transitory—it may perhaps altogether avoid it. But it is by no means evident that it will attract the funds just from where the gold would tend to flow, and it may well be that it only passes on the necessity of credit contraction to another country. And if for some reason all or the majority of central banks should at a particular moment feel that they ought to become more liquid and for this purpose raise their discount rates, the sole effect will be a kind of general tug-of-war in which all central banks, trying to stop an outflow of funds and if possible to attract funds, only succeed in bringing about a violent contraction of credit at home. But although the fact that central banks react to all major gold movements with changes in the rate of discount may mean that changes in the volume and direction of short term credits will be more frequent and violent if we have a number of banking systems organized on national lines, it is again not the fact that the system is international, but rather that is creates impediments ot the free international flow of funds which must be regarded as responsible for these disturbances.

Again we must be careful not to ascribe this difficulty

[1] If this effect was disregarded in the discussion of changes in the discount rates in the two preceeding lectures, this was done to make the effects discussed there stand out more clearly; but this must not be taken to mean that this effect on capital movements is not, at any rate in the short run, perhaps the most important effect of these changes.

to the existence of central banks in particular, although in a sense the growth of the sort of credit structure to which they are due was only made possibe by the existence of some such institutions. The ultimate source of the difficulty is the differentiation between moneys of different degrees of acceptability or liquidity, the existence of a structure consisting of superimposed layers of reserves of different degrees of liquidity, which makes the movement of short term money rates, and in consequence the movement of short term funds, much more dependent on the liquidity position of the different financial institutions than on changes in the demand for capital for real investment. It is because with " mixed " national monetary systems the movements of short term funds are frequently due, not to changes in the demand for capital for investment, but to changes in the demand for cash as liquidity reserves, that short term international capital movements have their bad reputation as causes of monetary disturbances. And this reputation is not altogether undeserved.

But now the question arises whether this defect can be removed not by making the medium of circulation in the different countries more homogeneous, but rather, as the Monetary Nationalists wish, by severing even the remaining tie between the national currencies, the fixed parities between them. This question is of particular importance since the idea that the national monetary authorities should never be forced by an outflow of capital to take any action which might unfavourably affect economic activity at home is probably the main source of the demand for variable exchanges. To this question therefore we must now turn.

4

The chief questions which we shall have to consider here are three : will the volume of short term capital movement be larger or smaller when there exists uncertainty about the future of exchange rates ? Are the national monetary authorities in a position either to prevent capital movements which they regard as undesirable, or to offset their effects ? And, finally, what further measures, if any, are necessary if the aims of such a policy are to be consistently followed ?

We have already partly furnished the answer to the first question. Although the contrary has actually been asserted, I am altogether unable to see why under a regime of variable exchanges the volume of short term capital movements as well as the frequency of changes in their direction should be anything but greater.[1] Every suspicion that exchange rates were likely to change in the near future would create an additional powerful motive for shifting funds from the country whose currency was

[1] The only argument against this view which I find at all intelligible is that, under the gold standard, movements to one of the gold points will create a certain expectation that the movement will soon be reversed and thus provides a special inducement to speculative shifts of funds. But while this is perfectly true, it only shows that the defects of the traditional gold standard were due to the fact that it was not a homogeneous international currency. If the same arrangements applied to international as to infranational payments the problem would disappear. This would be the case either if within the country as much as between countries the costs of transfers of money were not borne by some institution like the central banks and consequently (as in the United States before the establishment of the Federal Reserve System) rates of exchange between the different towns were allowed to fluctuate, and if at the same time gold were freely obtainable near the frontier as well as in the capital, or on the other hand, if the system of par clearance were applied to international as well as national payments. On the last point compare below, lecture V, p. 84.

likely to fall or to the country whose currency was likely to rise. I should have thought that the experience of the whole post-war period and particularly of the last few years had so amply confirmed what one might have expected *a priori* that there could be no reasonable doubt about this.[1] There is only one point which perhaps still deserves to be stressed a little further. Where the possible fluctuations of exchange rates are confined to narrow limits above and below a fixed point, as between the two gold points, the effect of short term capital movements will be on the whole to reduce the amplitude of the actual fluctuations, since every movement away from the fixed point will as a rule create the expectation that it will soon be reversed. That is, short term capital movements will on the whole tend to relieve the strain set up by the original cause of a temporarily adverse balance of payments. If exchanges, however, are variable, the capital movements will tend to work in the same direction as the original cause and thereby to intensify it. This means that if this original cause is already a short term capital movement, the variability of exchanges will tend to multiply its magnitude and may turn what originally might have been a minor inconvenience into a major disturbance.

Much more difficult is the answer to the second question : can the authorities control these movements; since what the monetary authorities can achieve in a particular

---

[1] Since it is being more and more forgotten that the period before 1931 was, on pre-war standards, already one of marked instability — and uncertainty about the future — of exchange rates, it is perhaps worth stressing that in particular the accumulation of foreign balances in London during that period was almost entirely a consequence of the fact that Sterling was regarded as relatively the most safe of the European currencies. Cf. on this T. E. GREGORY, *The Gold Standard and its Future*, Third edition, 1934, pp. 48 *et seq.*

INTERNATIONAL CAPITAL MOVEMENTS 65

direction will largely depend on what other consequences of their action they are willing to put up with. In the particular case the question is mainly whether they would be willing to let exchange rates fluctuate to any degree or whether they would not feel that although moderate fluctuations of exchange rates were not worth the cost of preventing them, yet they must not be allowed to exceed certain limits, since the unsettling effects from large fluctuations would be worse than the measures by which they could be prevented. In practice we must probably assume that even if the authorities are prepared to allow a slow and gradual depreciation of exchanges, they would feel bound to take strong action to counteract it as soon as it threatened to lead to a flight of capital or a strong rise of prices of imported goods.

The theory that by keeping exchange rates flexible a country could prevent dear money abroad from affecting home conditions is of course not a new one. It was for instance argued by the opponents of the introduction of the gold standard in Austria in 1892 that the paper standard insulated and protected Austria from disturbances originating on the world markets. But I doubt whether it has ever been carried quite as far as by some of our contemporary Monetary Nationalists, for instance Mr. Harrod, who declared that he could not accept exchange stabilisation " if thereby a country is committed to an interior monetary policy which involves raising the bank rate of interest ".[1] The modern idea apparently is that never under any circumstances must an outflow of capital be allowed to raise interest rates at home, and the advocates of this view seem to be satisfied that if the central banks are not commited to maintain a particular parity

[1] *Cf. Report of the Proceedings of the Meeting of Economists held at the Antwerp Chamber of Commerce* on July the 11th, 12th, and 13th 1935, published by the Antwerp Chamber of Commerce, p. 107.

they will have no difficulty either in preventing an outflow of capital altogether or in offsetting its effect by substituting additional bank credit for the funds which have left the country.

It is not easy to see on what this confidence is founded. So long as the outward flow of capital is not effectively prevented by other means, a persistent effort to keep interest rates low can only have the effect of prolonging this tendency indefinitely and of bringing about a continuous and progressive fall of the exchanges. Whether the outward flow of capital starts with a withdrawal of balances held in the country by foreigners, or with an attempt on the parts of nationals of the country to acquire assets abroad, it will deprive banking institutions at home of funds which they were able to lend, and at the same time lower the exchanges. If the central bank succeeds in keeping interest rates low in the first instance by substituting new credits for the capital which has left the country, it will not only perpetuate the conditions under which the export of capital has been attractive; the effect of capital exports on the rates of exchange will, as we have seen, tend to become self-inflammatory and a " flight of capital " will set in. At the same time the rise of prices at home will increase the demand for loans because it means an increase in the " real " rate of profit. And the adverse balance of trade which must necessarily continue while part of the receipts from exports is used to repay loans or to make new loans abroad, means that the supply of real capital and therefore the " natural " or " equilibrium " rate of interest in the country will rise. It is clear that under such conditions the central bank could not, merely by keeping its discount rate low, prevent a rise of interest rates without at the same time bringing about a major inflation.

5

If this is correct it would be only consistent if the advocates of Monetary Nationalism should demand that monetary policy proper should be supplemented by a strict control of the export of capital. If the main purpose of monetary management is to prevent exports of capital from disturbing conditions of the money market at home, this clearly is a necessary complement of central banking policy. But those who favour such a course seem hardly to be conscious of what it involves. It would certainly not be sufficient in the long run merely to prohibit the more conspicous forms of sending money abroad. It is of course true that if there are no impediments to the export of capital the most convenient and therefore perhaps the quantitatively most important form which the export of capital will take is the actual transfer of money from country to country. And it is conceivable that this might be pretty effectively prevented by mere prohibition and control. To make even this really effective would of course involve not only a prohibition of foreign lending and of the import of securities of any description, but could hardly stop short of a full-fledged system of foreign exchange control. But exchange control designed to prevent effectively the outflow of capital would really have to involve a complete control of foreign trade, since of course any variation in the terms of credit on exports or imports means an international capital movement.

To anyone who doubts the importance of this factor, I strongly recommend the very interesting memorandum on International Short Term Indebtedness which has recently been published by Mr. F. G. Conolly of the staff of the Bank for International Settlements in the recent joint publication of the Carnegie Endowment and the

International Chamber of Commerce.[1] I will quote only one paragraph. " It has been the experience of every country whose currency has come under pressure ", writes Mr. Conolly, " that importers tend not only to refuse to utilise the normal period of credit but to cover their requirements for months in advance; they prefer to utilise the home currency while it retains its international value rather than run the risk of being forced to pay extra for the foreign currency necessary for their purchases. Exporters, on the other hand, tend to allow foreign currencies, the proceeds of exports already made, to lie abroad and to finance their current operations as far as possible by borrowing at home. Thus a double strain falls on the exchange market; the normal supply of foreign currencies from export dries up while the demands from importers greatly increase. For a country with a large foreign trade the strain on the exchange market due to the effects of this change over in trade financing may be very considerable. "[2]. What Mr. Conolly here describes amounts, of course, to an export of capital which could only be prevented by controlling the terms of every individual transaction of the country's foreign trade, an export of capital which may be equally formidable whether the country carries on its foreign trade " actively " or " passively ",[3] that is whether it normally provides the capital to finance the trade herself or she borrows it. Indeed to anyone who has had any

---

[1] *The Improvement of Commercial Relations Between Nations. The Problem of Monetary Stabilization.* Separate Memoranda from the Economists consulted by the Joint Committee of the Carnegie Endowment and the International Chamber of Commerce and practical Conclusions of the Expert Committee appointed by the Joint Committee. Paris, 1936, pp. 352 et seq.

[2] *Ibid.*, p. 360

[3] Cf. N. G. Pierson, *The Problem of Value in a Socialist Community. Collectivist Economic Planning*, London, 1935.

INTERNATIONAL CAPITAL MOVEMENTS 69

experience of foreign exchange control there should be no doubt possible that an export of capital can only be prevented by controlling not only the volume of exports and imports so that they will always balance, but also the terms of credit of all these transactions.

At first indeed, and so long as discrepancies between national rates of interest are not too big and people have not yet fully learnt to adapt themselves to fluctuating exchanges, much less thoroughgoing measures may be quite effective. I can already hear some of my English friends point out to me the marvellous discipline of the City of London which on a slight hint from the Bank that capital exports would be undesirable will refrain from acting against the general interest. But we need only visualize how big the discrepancies between national interest rates would become if capital movements were for a time effectively stopped in order to realize how illusionary must be the hopes that anything but the strictest control will be able to prevent them.

But let us disregard for the moment the technical difficulties inherent in any effective control of international capital movements. Let us assume that the monetary authorities are willing to go any distance in creating new impediments to international trade and that they actually succeed in preventing any unwanted change in international indebtedness. Will this successfully insulate a country against the shocks which may result from changes in the rates of interest abroad? Or will these not still transmit themselves *via* the effect such a change of interest rates will have on the relative prices of the internationally traded securities and commodities? It is probably obvious that so long as there is a fairly free international movement of securities no great divergence in the movement of rates of interest in the different countries can persist for any length of time. But Monetary

Nationalists would probably not hesitate at any rate to attempt to inhibit these movements. It is not so generally recognised however that commodity movements will have a similar effect and perhaps this needs a few more words of explanation.

It will probably not be denied that a considerable rise in the rate of interest will lead to a fall in the prices of some commodities relatively to those of others, particularly of those which are largely used for the production of capital goods and of those of which large stocks are held, compared with those which are destined for more or less immediate consumption. And surely, in the absence of immediate adjustments in tariffs or quotas, such a fall will transmit itself to the prices of similar commodities in the country in which interest rates at first are not allowed to rise. But if the prices of the goods which are largely used for investment fall relatively to the prices of other goods this means an increased profitability of investment compared with current production, consequently an increased demand for loans at the existing rates of interest, and, unless the central bank is willing to allow an indefinite expansion of credit, it will be compelled by the rise of interest rates abroad to raise its own rate of interest, even if any outflow of capital has been effectively prevented. Although the supply of capital may not change, the kind of goods which under the changed circumstances it will be most profitable to import and export will yet alter the demand for capital with the same effects.[1]

The truth of the whole matter is that for a country which is sharing in the advantages of the international division of labour it is not possible to escape from the

---

[1] Cf. on this L. ROBBINS, *The Great Depression*, London, 1934, p. 175.

effects of disturbances in these international trade relations by means short of severing all the trade ties which connect it with the rest of the world. It is of course true that the less the points of contact with the rest of the world the less will be the extent to which disturbances originating outside the country will affect its internal conditions. But it is an illusion that it would be possible, while remaining a member of the international commercial community, to prevent disturbances from the ouside world from reaching the country, by following a national monetary policy such as would be indicated if the country were a closed community. It is for this reason that the ideology of Monetary Nationalism has proved, and if it remains influential will prove to an oven greater extent in the future, to be one of the main forces destroying what remnants of an international economic system we still have.

There are two more points which I should like specially to emphasize before I conclude for to-day. One is that up to this point I have, following the practice of the Monetary Nationalists, considered mainly the disturbing effects on a country of changes in the demand for capital originating abroad. But there is of course an other side to this picture. What from the point of view of the country to which the effects are transmitted from abroad is a disturbance is from the point of view of the country where the original change takes place a stabilising effect. To have to give up capital because somewhere else a sudden more urgent demand has arisen is certainly unsettling. But to be able to obtain capital at short notice if a sudden unforeseen need arises at home will certainly tend to stabilise conditions at home. It is more than unlikely that fluctuations on the national capital market would be smaller if the world were cut up into watertight compartments. The probability is rather that in this case fluctuations within each national territory would be much more violent and disturbing than they are now.

Closely conected with this is the second point, on which I can only touch even more shortly. I have already mentioned the probability that the restrictions on capital movements involved in a policy of Monetary Nationalism would tend to increase the differences between national interest rates. This would of course be due to the fact that while instability of exchange rates would tend to increase the volume and frequency of irregular flows of short term funds, it would to an even greater degree decrease the volume of international long term investment. Although by some this is regarded as a good thing, I doubt whether they fully appreciate what it would mean. The purely ecenomic effects, the restriction of international division of labour which it implies, and the reduction in the total volume of investment to which it would almost certainly lead, are bad enough. But even more serious seem to me the political effects of the intensification of the differences in the standard of life between different countries to which it would lead. It does not need much imagination to visualize the new sources of international friction which such a situation would create.[1] But this leads me beyond the proper scope of these lectures and I must confine myself to drawing your attention to it without attempting to elaborate it any further.

---

[1] Cf. in this L. ROBBINS, *Economic Planning and International Order*, pp. 68 et seq.

Lecture V

THE PROBLEMS
OF A REALLY INTERNATIONAL STANDARD

1

I have now concluded the negative part of my argument, the case against independent national currencies. While I cannot hope in the space of these few lectures completely to have refuted the theoretical basis of Monetary Nationalism, I hope at least to have shown three things : that there is no rational basis for the separate regulation of the quantity of money in a national area which remains a part of a wider economic system; that the belief that by maintaining an independent national currency we can insulate a country against financial shocks originating abroad is largely illusory; and that a system of fluctuating exchanges would on the contrary introduce new and very serious disturbances of international stability. I do not want now further to add to this except that I might perhaps remind you that my argument throughout assumed that such a system would be run as intelligently as is humanly possible. I have refrained from supporting my case by pointing to the abuses to which such a system would almost certainly lend itself, to the practical impossibility of different countries agreeing on what degree of depreciation is justified, to the consequent danger of competitive depreciation, and the general return to mercantilist policies of restriction which now, as in earlier centuries, are the inevitable reaction to debasement in other countries.[1]

We must recognise, therefore, that independent regulation of the various national currencies cannot be regarded as in any sense a substitute for a rationally regulated world monetary system. Such a system may to-day seem an unattainable ideal. But this does not mean that the question of what we can do to get as near the ideal as may be practicable does not present a number of important problems. Of course some " international " systems would be far from ideal. I hope I have made it clear in particular that I do not regard the sort of international system which we have had in the past as by any means completely satisfactory. The Monetary Nationalists condemn it because it is international; I, on the other hand, ascribe its shortcomings to the fact that it is not international enough. But the question how we can make it more satisfactory, that is more genuinely international, I have not yet touched upon. It is a question which raises exceedingly difficult problems; I can survey them only rapidly in this final lecture.

The first, but by no means the most important or most interesting question which I must consider is the question whether the international standard need be gold. On purely economic grounds it must be said that there are hardly any arguments which can be advanced for, and many serious objections which can be raised against, the use of gold as the international money. In a securely established world State with a government immune

[1] I feel I must remind the reader here that limitations of time made it impossible for me to dwell in these lectures on the tremendously important practical effects of a policy of Monetary Nationalism on commercial policy as long as I should have wished. Although this is well trodden ground, it cannot be too often reiterated that without stability of exchange rates it is vain to hope for any reduction of trade barriers.

A REALLY INTERNATIONAL STANDARD 75

against the temptations of inflation it might be absurd to spend enormous effort in extracting gold out of the earth if cheap tokens would render the same service as gold with equal or greater efficiency. Yet in a world consisting of sovereign national States there seem to me to exist compelling political reasons why gold (or the precious metals) alone and no kind of artificial international currency, issued by some international authority, could be used successfully as the international money. It is essential for the working of an international standard that each country's holdings of the international money should represent for it a reserve of exchange medium which in all eventualities will remain universally acceptable in international transactions. And so long as there are separate sovereign States there will always loom large among these eventualities the danger of war, or of the breakdown of the international monetary arrangements for some other reason. And since people will always feel that against these emergencies they will have to hold some reserve of the one thing which by age-long custom civilized as well as uncivilized people are ready to accept —that is, since gold alone will serve one of the purposes for which stocks of money are held—and since to some extent gold will always be held for this purpose, there can be little doubt that it is the only sort of international standard which in the present world has any chance of surviving. But, to repeat, while an international standard is desirable on purely economic grounds, the choice of gold with all its undeniable defects is made necessary entirely by political considerations.

What should be done if the well-known defects of gold should make themselves too strongly felt, if violent changes in the condition of its production or the appearance of a large new demand for it should threaten sudden changes in its value, is of course a problem of major importance. But it is neither the most interesting nor the most important problem and I do not propose to discuss it here. The difficulties which I want to consider are rather those which were inherent in the international gold standard, even before 1914, and to a still greater degree during its short post-war existence. They are the problems which arise out of the fact that the so-called gold currencies are connected with gold only through the comparatively small national reserves which form the basis of a multiple superstructure of credit money which itself consists of many different layers of different degrees of liquidity or acceptability. It is, as we have seen, this fact which makes the effects of changes in the international flow of money different from merely interlocal shifts, to which is due the existence of separate national monetary systems which to some extent have a life of their own. The homogeneity of the circulating medium of different countries has been destroyed by the growth of separate banking systems organized on national lines. Can anything be done to restore it?

2

It is important here first to distinguish between the need for some " lender of last resort " and the organization of banking on the " national reserve " principle. That an extensive use of bank deposits as money would not be possible, that deposit banking of the modern type could not exist, unless somebody were in a position to provide the cash if the public should suddenly want to convert a considerable part of its holdings of bank deposits into more liquid forms of money, is probably beyond doubt. It is far less obvious why all the banking institutions in a particular area or country should be made to rely on a single national reserve. It is certainly not a system which anybody would have deliberately devised on rational grounds and it grew up as an accidental by-product of a policy concerned with different problems.[1] The rational choice would seem to lie between either a system of " free banking ", which not only gives all banks the right of note issue and at the same time makes it necessary for them to rely on their own reserves, but also leaves them free to choose their field of operation and their correspondents without regard to national boundaries,[2] and on the other hand, an international central bank. I need not add that both of these ideals seem utterly impracticable in the world as we know it. But I am not certain whether the compromise we have chosen, that of national central banks which have no direct power over the bulk of the national circulation but which hold as the sole ultimate reserve a comparatively small amount of gold is not one of the most unstable arrangements imaginable.

Let us recall for a moment the essential features of the so-called gold standard systems as they have existed in modern times. The most widely used medium of exchange, bank deposits, is not fixed in quantity. Additional deposits may at any time spontaneously spring up (be " created " by the banks) or part of the total may similarly disappear. But while they are predominantly used in actual payments, they are by no means the only forms in which balances can be held to meet such payments. In this function deposits on current account are only one item—a very liquid one, although by no means the most liquid of all—in a long range of assets of varying

---

[1] Cf. W. BAGEHOT, *Lombard Street*, and V. C. SMITH, *The Rationale of Central Banking*, London, 1936.
[2] Cf. L. V. MISES, *Geldwertstabilisierung und Konjunkturpolitik*, Jena, 1928.

degree of liquidity.¹ Overdraft facilities, saving deposits and many types of very marketable securities on the one hand, and bank notes and coin on the other, will at different times and to different degrees compete with bank deposits in this function. And the amounts which will be held on current account to meet expected demands need not therefore fluctuate with the expected magnitude of these payments; they may also change with any change in the views about the ease with which it will be possible to convert these other assets into bank deposits. The supply of bank deposits on the other hand will depend on similar considerations. How much the banks will be willing to owe in this form in excess of the ready cash they hold will depend on their view as to how easy it will be to convert other assets into cash. It is when general confidence is high, so that comparatively small amounts of bank deposits will be needed for a given volume of payments, that the banks will be more ready to increase the amount of bank deposits. On the other hand, any increase of uncertainty about the future will lead to an increased demand for all the more liquid forms of assets, that is, in particular, for bank deposits and cash, and to a decrease in the supply of bank deposits.

Where there is a central bank the responsibility for the provision of cash for the conversion of deposits is divided between the banks and the central bank, or one should probably better say shifted from the banks to the central bank, since it is now the recognized duty of the central banks to supply in an emergency—at a price—all the cash that may be needed to repay deposits. Yet while the ultimate responsibility to provide the cash when needed is thus placed on the central bank, until this demand actually arises, the latter has little power to pre-

[1] Cf. particularly J. R. HICKS, *A Suggestion for the Simplification of the Theory of Money* (*Economica*, N. S. vol. II/5, February 1935, and F. LAVINGTON, *The English Capital Market*, 1921, p. 30.

A REALLY INTERNATIONAL STANDARD 79

vent the expansion leading to an increased demand for cash.

But with an international standard a national central bank is itself not a free agent. Up to this point the cash about which I have been speaking is the money created by the central bank which within the country is generally acceptable and is the only means of payment outside the circle of the customers of a particular bank. The central bank, however, has not only to provide the required amounts of the medium generally accepted within the country; it has also to provide the even more liquid, internationally acceptable, money. This means that in a situation where there is a general tendency towards greater liquidity there will be at the same time a greater demand for central bank money and for the international money. But the only way in which the central bank can restrict the demand for and increase the supply of the international money is to curtail the supply of central bank money. In consequence, in this stage as in the preceding one, any increase in the demand for the more liquid type of money will lead to a much greater decrease in the supply of the somewhat less liquid kinds of money.

This differentiation between the different kinds of money into those which can be used only among the customers of a particular bank and those which can be used only within a particular country and finally those which can be used internationally—these artificial distinctions of liquidity (as I have previously called them)—has the effect, therefore, that any change in the relative demand for the different kinds of money will lead to a cumulative change in the total quantity of the circulating medium. Any demand on the banks for conversion of part of their deposits into cash will have the effect of compelling them to reduce their deposits by more than the amount paid out and to obtain more cash from the central bank, which in turn will be forced to take counter-measures and so to transmit the tendency towards contraction to the other banks. And the same applies, of course, *mutatis mutandis* to a decrease in the demand for the more liquid type of assets, which will bring about a considerable increase in the supply of money.

All this is of course only the familiar phenomenon which Mr. R. G. Hawtrey has so well described as the "inherent instability of credit". But there are two points about it which deserve special emphasis in this connection. One is that, in consequence of the particular organisation of our credit structure, changes in liquidity preference as between different kinds of money are probably a much more potent cause of disturbances than the changes in the preference for holding money *in general* and holding goods *in general* which have played such a great role in recent refinements of theory. The other is that this source of disturbance is likely to be much more serious when there is only a single bank for a whole region or when all the banks of a country have to rely on a single central bank; since the effect of any change in liquidity preference will generally be confined to the group of people who directly or indirectly rely on the same reserve of more liquid assets.

It seems to follow from all this that the problem with which we are concerned is not so much a problem of currency reform in the narrower sense as a problem of banking reform in general. The seat of the trouble is what has been very appropriately been called the perverse elasticity of bank deposits [1] as a medium of circulation, and the cause of this is that deposits, like other forms of "credit money", are claims for another, more

---

[1] L. CURRIE, *The Supply and Control of Money in the United States*, Cambridge, 1934, pp. 130 et seq.

A REALLY INTERNATIONAL STANDARD 81

generally acceptable sort of money, that a proportional reserve of that other money must be held against them, and that their supply is therefore inversely affected by the demand for the more liquid type of money.

3

By far the most interesting suggestion on Banking Reform which has been advanced in recent years, not because in its present form its seems to be practicable or even theoretically right, but because it goes to the heart of the problem, is the so-called Chicago or 100 per cent plan.[1] This proposal amounts in effect to an extension of the principles of Peel's Act of 1844 to bank deposits. The most praticable suggestion yet made for its execution is to give the banks a sufficient quantity of paper money to increase the reserves held against demand deposits to 100 per cent and henceforth to require them to maintain permanently such a 100 per cent reserve.

In this form the plan is conceived as an instrument of Monetary Nationalism. But there is no reason why it should not equally be used to create a homogeneous international currency.[2] A possible, although perhaps somewhat fantastic, solution would seem to be to reduce proportionately the gold equivalents of all the different national monetary units to such an extent that all the money in all countries could be covered 100 per cent by gold, and from that date onwards to allow variations in the national circulations only in proportion to changes

---

[1] On the significance of the " Chicago Plan " compare particularly the interesting and stimulating article by H. C. SIMONS, *Rule versus Authority in Monetary Policy* (*Journal of Political Economy*, vol. 44, no. 1, February 1936, and F. LUTZ, *Das Grundproblem der Geldverfassung*, 1936, where references to the further literature on the proposal will be found.

[2] Cf. H. C. SIMONS, *loc. cit.*, p. 5, note 3.

in the quantity of gold in the country.[1] Such a plan would clearly require as an essential complement an international control of the production of gold, since the increase in the value of gold would otherwise bring about an enormous increase in the supply of gold. But this would only provide a safety valve probably necessary in any case to prevent the system from becoming all too rigid

The undeniable attractivenes of this proposal lies exactly in the feature which makes it appear somewhat impracticable, in the fact that in effect it amounts, as is fully realized by at least one of its sponsors, to an abolition of deposit banking as we know.[2] It does provide, instead of the variety of media of circulation which to-day range according to their degree of acceptabillty from bank deposits to gold, one single kind of money. And it would do away effectively with that most pernicious feature of our present system : namely that a movement towards more liquid types of money causes an actual decrease in the total supply of money and *vice versa*. The most serious question which it raises, however, is whether by abolishing deposit banking as we know it we would effectively prevent the principle on which it rests from manifesting itself in other forms. It has been well remarked by the most critical among the originators of the scheme that banking is a pervasive phenomenon[3] and the question is whether, when we prevent it from appearing in its traditional form, we will not just drive it into other and less easily controllable forms. Historical precedent rather suggests that we must be wary in this respect. The Act of 1844 was designed to control what then seemed to be the only important substitute for gold as a widely used

---

[1] A perhaps somewhat less impracticable alternative might be international bimetallism at a suitable ratio.
[2] Cf. H. C. SIMONS, *loc. cit.*, p. 16.
[3] *Ibid.*, p. 17.

A REALLY INTERNATIONAL STANDARD 83

medium of exchange and yet failed completely in its intention because of the rapid growth of bank deposits. Is it not possible that if similar restrictions to those placed on bank notes were now placed on the expansion of bank deposits, new forms of money substitutes would rapidly spring up or existing ones would assume increasing importance? And can we even to-day draw a sharp line between what is money and what is not? Are there not already all sorts of " near-moneys "[1] like saving deposits, overdraft facilities, bills of exchange, etc., which satisfy at any rate the demand for liquid reserves nearly as well as money?

I am afraid all this must be admitted, and it considerably detracts from the alluring simplicity of the 100 per cent banking scheme. It appears that for this reason it has now also been abandoned by at least one of its original sponsors.[2] The problem is evidently a much wider one and I agree with Mr. H. C. Simons that it " cannot be dealt with merely by legislation directed at what we call banks ".[3] Yet in one respect at least the 100 per cent proposal seems to me to point in the right direction. Even if, as is probably the case, it is impossible to draw a sharp line between what is to be treated as money and what is not, and if consequently any attempt to fix rigidly the quantity of what is more or less arbitrarily segregated as " money " would create serious difficulties, it yet remains true that, within the field of instruments which are undoubtedly generally used as money, there are unnecessary and purely institutional distinctions of liquidity which are the sources of serious disturbances and which should as far as possible be eliminated. If this cannot be done for the time being by a general return to

---
[1] *Op. cit.*, p. 17.
[2] *Ibid.*, p. 17.
[3] *Ibid.*

the common use of the same international medium in the great majority of transactions, it should at least be possible to approach this goal by reducing the distinctions of liquidity between the different kinds of money actually used, and offsetting as far as possible the effects of changes in the demand for liquid assets on the total quantity of the circulating medium.

4

This brings me to the more practical question of what can be done to diminish the instability of the credit structure if the general framework of the present monetary system is to be maintained. The aim, as we have just seen, must be to increase the certainty that one form of money will always be readily exchangeable against other forms of money at a known rate, and that such changes should not lead to changes in the total quantity of money. In so far as the relations between different national currencies are concerned this leads, of course, to a demand for reforms in exactly the opposite direction from those advocated by Monetary Nationalists. Instead of flexible parities or a widening of the " gold points ", absolute fixity of the exchange rates should be secured by a system of international par clearance. If all the central banks undertook to buy and sell foreign exchange freely at the same fixed rates, and in this way prevented even fluctuations within the " gold points ", the remaining differences in denomination of the national currencies would really be no more significant than the fact that the same quantity of cloth can be stated in yards and in meters. With an international gold settlement fund on the lines of that operated by the Federal Reserve System, which would make it possible to dispense with the greater part of the actual gold movements which used to take place in

A REALLY INTERNATIONAL STANDARD 85

the past, invariable rates of exchange could be maintained without placing any excessive burden on the central banks.[1] The main aim here would of course be rather to remove one of the main causes of international movements of short term funds than to prevent such movements or to offset their effects by means which will only increase the inducement to such movements.[2]

But invariability of the exchange rates is only one precondition of a successful policy directed to minimize monetary disturbances. It eliminates one of the institutional differentiations of liquidity which are likely to give rise to sudden changes in favour of holding one sort of money instead of another. But there remains the further distinction between the different sorts of money which constitute the national monetary systems; and, so long as the general framework of our present banking systems is retained, the dangers to stability which arise here can hardly be combatted otherwise than by a deliberate policy of the national central banks.

The most important change which seems to be neces-

---

[1] The founders of the Bank for International Settlements definitely contemplated that the Bank might establish such a fund and article 24 of it Statutes specifically states that the bank may enter into special agreements with central banks to facilitate the settlement of international transactions between them. — " For this purpose it may arrange with central banks to have gold earmarked for their account and transferable on their order, to open accounts through which central banks can transfer their assets from one currency to another and to take such other measures as the Board may think advisable within the limits of the powers granted by these Statutes ".

[2] In a book which has appeared since these lectures were delivered (C. R. WHITTLESEY, *International Monetary Issues*, New York, 1937) the author, after pointing out that a widening of the gold points would have the effects of increasing the volume of short term capital movements of this sort (p. 116) concludes that " the only way of overcoming this factor would be to eliminate the gold points " (p. 117). But the only way of eliminating the gold points of which he can think is to abolish the gold standard!

sary here is that the gold reserves of all the central banks should be made large enough to relieve them of the necessity of bringing about a change in the total national circulation *in proportion* to the changes in their reserves; that is, that any change in the relative amounts of money in different countries should be brought about by the actual transfer of corresponding amounts from country to country without any " secondary " contractions and expansions of the credit super-structure of the countries concerned. This would be the case only if individual central banks held gold reserves large enough to be used freely without resort to any special measures for their " protection ".

Now the present abundance of gold offers an exceptional opportunity for such a reform. But to achieve the desired result not only the absolute supply of gold but also its distribution is of importance. In this respect it must appear unfortunate that those countries which command already abundant gold reserves and would therefore be in a position to work the gold standard on these lines, should use that position to keep the price artificially high. The policy on the part of those countries which are already in a strong gold position, if it aims at the restoration of an international gold standard, should have been, while maintaining constant rates of exchange with all countries in a similar position, to reduce the price of gold in order to direct the stream of gold to those countries which are not yet in a position to resume gold payments. Only when the price of gold had fallen sufficiently to enable those countries to acquire sufficient reserves should a general and simultaneous return to a free gold standard be attempted.

It may seem at first that even if one could start with an appropriate distribution of gold between countries which at first would put each country in a position where

A REALLY INTERNATIONAL STANDARD   87

it could allow its stock of gold to vary by the absolute amounts by which its circulation would have to increase or decrease, some countries would soon again find their gold stocks so depleted that they would be compelled to take traditional measures for their protection. And it cannot be denied that so long as the stock of gold of any country is anything less than 100 per cent of its total circulation, it is at least conceivable that it may be reduced to a point were in order to protect the remainder the monetary authorities might have to have recourse to a policy of credit contraction. But a short reflection will show that this is extraordinarily unlikely to happen if a country starts out with a fairly large stock of gold and if its monetary authorities adhere to the main principle not only with regard to decreases but equally with regard to increases in the total circulation.

If we assume the different countries to start with a gold reserve amounting to only a third of the total monetary circulation [1] this would probably provide a margin amply sufficient for any reduction of the country's share in the world's stock of money which is likely to become necessary. That a country's share in the world's income, and therefore its relative demand for money, should fall off by more than this would at any rate be an exceptional case requiring exceptional treatment. [2] If history seems

---

[1] At the present value of gold the world's stock of monetary gold (at the end of 1936) amounts to 73.5 per cent. of all sight liabilities of the central banks *plus* the circulation of Government paper money. The percentage would of course be considerably lower if, as would be necessary for this purpose, the comparison were made with the total of sight deposits with commercial banks *plus* bank notes etc. in the hands of the public. But there can be no doubt that even if the price of gold should be somewhat lowered (say by one seventh, *i. e.* from 140 to 120 shillings or from $ 35 to $ 30 per ounce) there would still be ample gold available to provide sufficient reserves.

[2] If in spite of this in an individual case the gold reserves of a country should be nearly exhausted, the necessary remedy would

88   A REALLY INTERNATIONAL STANDARD

to suggest that such considerable losses of gold are not at all infrequent, this is due to the operation of a different cause which should be absent if the principle suggested were really applied. If under the traditional gold standard any one country expanded credit out of step with the rest of the world this did usually bring about an outflow of gold only after a considerable time lag. This in itself would mean that, before equilibrium would be restored by the direct operations of the gold flows, an amount of gold approximately equal to the credit created in excess would have to flow out of the country. If, however, as has often been the case, the country should be tardy in decreasing its circulation by the amount of gold it has lost, that is, if it should try to " offset " the losses of gold by new creations of credit, there would be no limits to the amount of gold which may leave the country except the size of the reserves. Or in other words, if the principle of changing the total circulation by the full amount of gold imported or exported were strictly applied, gold movements would be much smaller than has been the case in the past, and the size of the gold movements experienced in the past create therefore no presumption that they would be equally large in the future.

5

These considerations will already have made it clear that the principle of central banking policy here proposed by

be to acquire the necessary amount of gold through an external loan and to give this amount to the central bank in repayment of part of the state debt which presumably will constitute at least part of its non gold assets (or in payments of any other assets which the bank would have to sell to the Government). The main point here is that the acquisition of this gold must be paid for out of taxation and not by the creation of additional credit by the central bank.

A REALLY INTERNATIONAL STANDARD 89

no means implies that the central banks should be relieved from all necessity of shaping their credit policy according to the state of their reserves. Quite the contrary. It only means that they should not be compelled to adhere to the mechanical rule of changing their notes and deposits *in proportion* to the change in their reserves. Instead of this they would have to undertake the much more difficult task of influencing the total volume of money in their countries in such a way that this total would change by the same absolute amounts as their reserves. And since the central bank has no direct power over the greater part of the circulating medium of the country it would have to try to control its volume indirectly. This means that it would have to use its power to change the volume of its notes and deposits so as to make the superstructure of credit built on those move in conformity with its reserves. But as the amount of ordinary bank deposits and other forms of common means of exchange based on a given volume of central bank money will be different at different times, this means that the central bank, in order to make the total amount of money move with its reserves, would frequently have to change the amount of central bank money independently of changes in its reserves and occasionally even in a direction opposite to that in which its reserves may change.

It should perhaps always have been evident that, with a banking system which has grown up to rely on the assistance of a central bank for the supply of cash when needed, no sort of control of the circulating medium can be achieved unless the central bank has power and uses this power to control the volume of bank deposits in ordinary times. And the policy to make this control effective will have to be very different from the policy of a bank which is concerned merely with its own liquidity. It will have to act persistently against the trend of the movement of credit in the country, to contract the credit basis when the superstructure tends to expand and to expand the former when the latter tends to contract.

It is to-day almost a commonplace that, with a developped banking structure, the policy of the central bank can in no way be automatic. It would indeed require the greatest art and discernment for a central bank to succeed in making the credit money provided by the private banks behave as a purely metallic circulation would behave under similar circumstances. But while it may appear very doubtful whether this ideal will ever be fully achieved, there can be no doubt that we are still so far from it that very considerable changes from traditional policy would be required before we shall be able to say that even what is possible has been achieved.

In any case it should be obvious that the existence of a central bank which does nothing to counteract the expansions of banking credit made possible by its existence only adds another link in the chain through which the cumulative expansions and contractions of credit operate. So long as central banks are regarded, and regard themselves, only as " lenders of last resort " which have to provide the cash which becomes necessary in consequence of a previous credit expansion with which, until this point arrives, they are not concerned, so long as central banks wait until " the market is in the Bank " before they feel bound to check expansion, we cannot hope that wide fluctuations in the volume of credit will be avoided. Certainly Mr. Hawtrey was right with his now celebrated statement that " so long as the credit is regulated with reference to reserve proportions, the trade cycle is bound to recur ".[1] But I am afraid only one and that not the more important of the essential corrolaries

---

[1] *Monetary Reconstruction*, London, 1923, p. 144.

A REALLY INTERNATIONAL STANDARD   91

of this proposition is usually derived from this statement. What is usually emphasized is the fact that concern with reserve proportions will ultimately compel central banks to stop a process of credit expansion and actually to bring about a process of credit contraction. What seems to me much more important is that sole regard to their own reserve proportions will not lead central banks to counteract the increase of bank deposits, even if it means an increase of the credit circulation of the country relatively to the gold reserve, and although it is an increase largely made possible by the certain expectation on the part of the other banks that the central bank will in the end supply the cash needed.

On the question how far central banks are in practice likely to succeed in this difficult task different opinions are clearly possible. The optimist will be convinced that they will be able to do much more than merely offset the dangers which their existence creates. The pessimist will be sceptical whether on balance they will not do more harm than good. The difficulty of the task, the impossibility of prescribing any fixed rule, and the extent to which the action of the central banks will always be exposed to the pressure of public opinion and political influence certainly justify grave doubts. And though the alternative solution is to-day probably outside of the realm of practical politics, it is sufficiently important to deserve at least a passing consideration before we leave this subject.

As I have pointed out before, the " national reserve principle " is not insolubly bound up with the centralization of the note issue. While we must probably take it for granted that the issue of notes will remain reserved to one or a few privileged institutions, these institutions need not necessarily be the keepers of the national reserve. There is no reason why the Banks of Issue should not be entirely confined to the functions of the issue department of the Bank of England, that is to the conversion of gold into notes and notes into gold, while the duty of holding appropriate reserves is left to individual banks. There could still be in the background—for the case of a run on the banks—the power of a temporary "suspension" of the limitations of the note issue and of the issue of an emergency currency at a penalizing rate of interest.

The advantage of such a plan would be that one tier in the pyramid of credit would be eliminated and the cumulative effects of changes in liquidity preference accordingly reduced. The disadvantage would be that the remaining competing institutions would inevitable have to act on the proportional reserve principle and that nobody would be in a position, by a deliberate policy, to offset the tendency to cumulative changes. This might not be so serious if there were numerous small banks whose spheres of operation freely overlapped over the whole world. But it can hardly be recommended where we have to deal with the existing banking systems which consist of a few large institutions covering the same field of a single nation. It is probably one of the ideals which might be practical in a liberal world federation but which is impracticable where national frontiers also mean boundaries to the normal activities of banking institutions. The practical problem remains that of the appropriate policy of national central banks.

6

It is unfortunately impossible to say here more about the principles which a rational central banking policy would have to follow without going into some of the most controversial problems of the theory of the trade cycle which clearly fall outside the scope of these lec-

A REALLY INTERNATIONAL STANDARD 93

tures. I must therefore confine myself to pointing out that what I have said so far is altogether independent of the particular views on this subject for which I have been accused, I think unjustly, of being a deflationist. Whether we think that the ideal would be a more or less constant volume of the monetary circulation, or whether we think that this volume should gradually increase at a fairly constant rate as productivity increases, the problem of how to prevent the credit structure in any country from running away in either direction remains the same.

Here my aim has merely been to show that whatever our views about the desirable behaviour of the total quantity of money, they can never legitimately be applied to the situation of a single country which is part of an international economic system, and that any attempt to do so is likely in the long run and for the world as a whole to be an additional source of instability. This means of course that a really rational monetary policy could be carried out only by an international monetary authority, or at any rate by the closest cooperation of the national authorities and with the common aim of making the circulation of each country behave as nearly as possible as if it were part of an intelligently regulated international system.

But I think it also means that so long as an effective international monetary authority remains an utopian dream, any mechanical principle (such as the gold standard) which at least secures some conformity of monetary changes in the national area to what would happen under a truly international monetary system is far preferable to numerous independent and independently regulated national currencies. If it does not provide a really rational regulation of the quantity of money, it at any rate tends to make it behave on roughly foreseeable lines, which is of the greatest importance. And since there is no means, short of complete autarchy, of protecting a country against the folly or perversity of the monetary policy of other countries, the only hope of avoiding serious disturbances is to submit to some common rules, even if they are by no means ideal, in order to induce other countries to follow a similarly reasonable policy. That there is much scope for an improvement of the rules of the game which were supposed to exist in the past, nobody will deny. The most important step in this direction is that the *rationale* of an international standard and the true sources of the instability of our present system should be properly appreciated. It was for this reason that I felt that my most urgent task was to restate the broader theoretical considerations which bear on the practical problem before us. I hope that by confining myself largely to these theoretical problems I have not too much disappointed the expectations to which the title of these lectures may have given rise. But, as I said at the beginning of these lectures, I do believe that in the long run human affairs are guided by intellectual forces. It is this belief which for me gives abstract considerations of this sort their importance, however slight may be their bearing on what is practicable in the immediate future.

Lightning Source UK Ltd.
Milton Keynes UK
UKHW01f1951160518
322730UK00003B/238/P